Rivers in History

HISTORY OF THE URBAN ENVIRONMENT

Martin V. Melosi and Joel A. Tarr, *Editors*

PERSPECTIVES ON WATERWAYS
IN EUROPE AND NORTH AMERICA

Edited by
Christof Mauch
and
Thomas Zeller

University of Pittsburgh Press

Published by the University of Pittsburgh Press, Pittsburgh, Pa., 15260
Copyright © 2008, University of Pittsburgh Press
All rights reserved
Manufactured in the United States of America
Printed on acid-free paper
10 9 8 7 6 5 4 3 2 1

Library of Congress Cataloging-in-Publication Data

Perspectives on waterways in Europe and North America / edited by Christof Mauch and Thomas Zeller.
  p. cm. — (History of the urban environment)
  Includes bibliographical references and index.
  ISBN-13: 978-0-8229-4345-7 (cloth : alk. paper)
  ISBN-10: 0-8229-4345-X (cloth : alk. paper)
  ISBN-13: 978-0-8229-5988-5 (pbk. : alk. paper)
  ISBN-10: 0-8229-5988-7 (pbk. : alk. paper)
  1. Rivers—Europe. 2. Rivers—North America. I. Mauch, Christof. II. Zeller, Thomas, 1966-
  GB1281.P47 2008
  333.91'62094—dc22
                                                    2008015489

Contents

	List of Illustrations	vii
	Acknowledgments	ix
1.	Rivers in History and Historiography: An Introduction **Christof Mauch and Thomas Zeller**	1
2.	"Time is a violent torrent": Constructing and Reconstructing Rivers in Modern German History **David Blackbourn**	11
3.	From Parisian River to National Waterway: The Social Functions of the Seine, 1750–1850 **Isabelle Backouche**	26
4.	Pittsburgh's Three Rivers: From Industrial Infrastructure to Environmental Asset **Timothy M. Collins, Edward K. Muller, and Joel A. Tarr**	41
5.	The Cultural and Hydrological Development of the Mississippi and Volga Rivers **Dorothy Zeisler-Vralsted**	63
6.	River Diking and Reclamation in the Alpine Piedmont: The Case of the Isère **Jacky Girel**	78
7.	Holding the Line: Pollution, Power, and Rivers in Yorkshire and the Ruhr, 1850–1990 **Charles E. Closmann**	89
8.	Saving the Rhine: Water, Ecology, and *Heimat* in Post–World War II Germany **Thomas Lekan**	110

CONTENTS

9. Postwar Perceptions of German Rivers: A Study of the Lech as Energy Source, Nature Preserve, and Tourist Attraction 137
   Ute Hasenöhrl

10. Viewing the Gilded Age River: Photography and Tourism along the Wisconsin Dells 149
    Steven Hoelscher

Notes 173
Contributors 219
Index 223

Illustrations

Figure 3.1. Map of Parisian ports and bridges along the Seine 29
Figure 3.2. Map of the Seine and canals in and around Paris before and after 1860 31
Figure 4.1. Pittsburgh's Mon Wharf (c. 1875) 45
Figure 4.2. Coal barges on the Ohio River (c. 1905) 45
Figure 4.3. Lock 2 of the Davis lock-and-dam system on the Ohio River 46
Figure 4.4. Pittsburgh industrial waterfronts (1957) 54
Figure 4.5. The peninsula at the confluence of Pittsburgh's three rivers, known as the Point 61
Figure 6.1. Hydrosystem of the Upper Rhône and Isère River valleys 79
Figure 6.2. Francesco-Luigi Garella's map of the Arc-Isère junction (1786–1789) 80
Figure 6.3. An engineering design for dikes along the Isère River (c. 1828) 84
Figure 6.4. A topographical plan of the warping system used in the Combe de Savoie (1846) 85
Figure 7.1. Map of rivers in Yorkshire, a county in the north of England 91
Figure 7.2. Map of waterways in the Ruhr region of Germany 99
Figure 8.1. Magazine graphic showing sources of pollution along the Rhine River (1986) 133
Figure 8.2. Two graphics from a magazine article on German waterways (1959) 134
Figure 10.1. "H. H. Bennett at Sugar Bowl," unknown photographer 152
Figure 10.2. "Sugar Bowl," H. H. Bennett 153
Figure 10.3. "Out of Boat Cave," H. H. Bennett 156
Figure 10.4. "Rowboat in Front of Eaton Grotto," H. H. Bennett 157
Figure 10.5. "Foot of High Rock," H. H. Bennett 158
Figure 10.6. "Berry's Landing," H. H. Bennett 159

Figure 10.7. "People on top of Stand Rock," H. H. Bennett — 161
Figure 10.8. *Map of the Dells of the Wisconsin River near Kilbourn City, Wisconsin*, Charles Lapham — 162
Figure 10.9. "Muscular Vigor in Repose" (*Raftsman's* Series), H. H. Bennett — 164
Figure 10.10. "We Are Broke Up—Take Our Line" (*Raftsman's* Series), H. H. Bennett — 165
Figure 10.11. "Lumber Rafts in Narrows," H. H. Bennett — 166
Figure 10.12. "Apollo No. 1 Steamboat in Cold Water Canyon," H. H. Bennett — 167
Figure 10.13. *Dells of the Wisconsin River at Kilbourn City*, postcard with photograph by H. H. Bennett — 169

Acknowledgments

The editors wish to thank the staff of the German Historical Institute in Washington, D.C., particularly Christa Brown and Bärbel Thomas, for their invaluable help in organizing the conference "Rivers in History: Designing and Conceiving Waterways in Europe and North America" in December 2003. Several of the papers presented at that conference formed the basis of this collection, and others that are not included here had a salutary impact on the contributions to this volume. The constructive criticism from Meredith McKittrick, Harold Platt, and Martin Reuss, who served as commentators during the conference, provided great inspiration. Likewise, the reviews from two anonymous readers did much to strengthen the individual chapters and the introduction. We owe a debt of gratitude to Mary Tonkinson at the German Historical Institute, whose skilled editing of the manuscript improved every one of its pages. It has also been a genuine pleasure to work with director Cynthia Miller, managing editor Deborah Meade, and their colleagues at the University of Pittsburgh Press, who embraced the idea for this volume with enthusiasm and shepherded it into print with such professionalism.

Munich/College Park, 2007

1

Rivers in History and Historiography
An Introduction

CHRISTOF MAUCH AND THOMAS ZELLER

Sources of both abundance and destruction, life and death, rivers have always had a powerful hold over humankind. They run through every human landscape, whether mythical or actual. In the Book of Genesis, the geography of humanity's first home is defined by a river that flows through Eden and separates into four headwaters, creating the Pishon, Gihon, Tigris, and Euphrates rivers. According to classical mythology, the boundaries of the underworld are likewise demarcated by rivers: the Acheron, Cocytus, Phlegethon, Lethe, Eridanos, and of course the Styx. Even the Epic of Gilgamesh (c. 2100 BCE) tells of a catastrophic river flood sent by angry deities to destroy all life.

As every anthropologist knows, the rise of civilizations has always been inextricably linked to the successful management of water when there was either too little or too much of it. Lewis Mumford has observed that "all the great historic cultures ... have thriven through the movement of men and institutions and inventions and goods along the natural highway of a great river," and over the centuries rivers have often become identified with the societies they supported.[1] Can one think of China without imagining the Yangzi, of ancient Egypt without recalling the Nile, of Caesar's Rome or Dante's Florence without picturing the Tiber or the Arno? Many writers of the past have claimed affinities between rivers and the communities connected to them. The Rhine and its people were said to be romantic; the Thames, imperial; the Rhône, savage; and so on.[2] This anthropomorphizing tendency often appears in patriotic and even jingoistic contexts; the qualities of particular rivers are alleged to reflect those of particular nations. The control of water and the ocean has been deeply inscribed into the perception of the Netherlands, for example, among both those inside and those outside the country. When a river became the focus of competing claims, as with the Rhine, then this essentialist link between the river's imagined attributes and regional character was all the more crucial to nationalistic myths.[3] Some national narratives have also described the transformation of existing rivers and the construction of artificial ones (in the form of canals) as feats that could be achieved only by powerful states, thus glorifying empires and extending their economic reach. Such was the case with China's Grand Canal, the oldest parts of which date back to the fifth century BCE.

Yet despite the power ascribed to them, in historical narratives rivers have typically been treated as a neutral setting rather than a dynamic force regarded as merely the backdrop against which human history unfolded.[4] Only in recent decades have historians begun to pay attention to rivers themselves, addressing the topic of environmental change in waterways and building on the work of geomorphologists and biologists who have studied the human impact on rivers over time—impacts that may be direct (within the river channel) or indirect (outside the channel).[5] Very often, environmental changes occurring at some distance from a river have modified not only its water quality but even the course of the river itself. The rise of the fur trade in the American West, for example, which resulted in the killing of hundreds of thousands of beavers, rapidly affected regional waterways. By drastically reducing the number of beaver dams, the fur trade increased sediment transport, which, over time, changed the paths of many rivers and streams. Timber harvests in mountainous regions have similarly altered the courses of rivers around the globe—from the Alps to the Andes, from the Urals to the Rockies. In Europe and North America the construction of roads and urban development have had an enormous impact on individual rivers; building a road or a town affects vegetation and the movement of topsoil from the surrounding terrain and, eventually, a river's flow.[6]

What is more, the shape of many European and North American rivers has changed dramatically over the past century and a half. While early modern science had envisioned and begun to sustain a discipline of hydrology, it was only in the nineteenth century that rivers in these parts of the world were radically transformed by experts acting on behalf of centralizing nation-states. First and foremost was the continued use of rivers for transportation. Several European countries and the United States also engaged in a frenzy of canal building similar to the engineering euphoria that resulted in the construction of national road networks in the mid-twentieth century. The 240-kilometer Canal du Midi extended the Atlantic rivers of France to the Mediterranean by 1681; the canal craze in Great Britain transformed the English landscape during the second half of the eighteenth century; and in the United States some 4,400 miles of artificial waterways had been built by 1830. Because canals came to be seen as the ideal means of transporting goods, rivers were increasingly engineered to resemble these artificial waterways.[7] In order to accommodate growing cargo loads and larger barges, river channels were standardized in width and depth, which required dredging on a scale previously unseen.

In addition, riverbeds were straightened or, as the experts' jargon of the time would have it, "corrected." The goal of these undertakings, achieved with varying success, was flood prevention and management. While relatively undisturbed rivers will change their lateral course in response to the amount of water they carry, hydrologists and engineers of the nineteenth century sought to replace such vacillations with smooth, predictable, and, in effect, shorter rivers. One of these engineers, Johann Gottfried Tulla of the southwestern German duchy of Baden, used martial analogies to describe his work. For him the malarious Rhine was the enemy to be conquered and pacified.[8] While such metaphors are now eschewed, the most important institution for river design and management in the United States today is still a branch of the military, namely the Army Corps of Engineers.[9] Since the Civil War, the Corps has rendered some twenty-six thousand miles of waterways navigable to vessels drawing nine feet, thus turning the United States into one of the world's most extensive hydrological systems.[10]

The centerpiece of this system is, of course, the Mississippi, the catchment area of which covers some 40 percent of the continental United States. Nineteenth-century observers thought of it as "nature's highway to market," an artery for midwestern agricultural products and southern cotton to be shipped out of New Orleans. As early as the 1830s, the federal government began to remake the river for transportation purposes. Shoals and sandbars were removed, rocks and rapids were dynamited to provide clear passage, and meandering sloughs and backwaters were closed off to confine water flow to the main channel. From 1878 to 1930, the U.S. Congress authorized three major navigation projects for the Mississippi that ultimately produced a nine-foot navigation channel. Dozens of locks and dams became necessary components of this vast technological system. Such structures represent both the best and the worst of public works; while they eliminate environmental impediments to commerce and settlement, which promotes the circulation of freight, they also decrease ecological diversity and increase the risk of flooding as a result.[11] The devastating force of the hurricanes of summer 2005 on the Gulf Coast dramatically highlighted the weakness of some human and mechanical elements of this system. More than simply the breaching of levees, Hurricanes Katrina and Rita signified a systemic failure. The ineptitude of local, regional, and federal governments brought the existence of this hydrological system to the forefront of public debate—but only for a brief time.[12]

River dams are perhaps the most conspicuous features of modern river management. When the World Commission on Dams surveyed the globe in 2000, it counted more than forty-five thousand large dams. While most beavers and some humans have been damming rivers as long as either species has existed, the scope and scale of dam building in the twentieth century was unprecedented. The U.S. Bureau of Reclamation, the federal agency with oversight for irrigation projects in the American West, became an ardent proponent of professional dam building under central control. The dams were constructed to harness and distribute water and generate energy, but they had symbolic functions as well. The bureau's mega-project, the Hoover Dam outside Las Vegas, has been glorifying the nation and boosting the electricity grid's capacity since 1937.[13] This iconic structure became the prototype of a worldwide boom in dam building that started in the 1930s. While access to water, flood control, and electricity generation, by themselves or in combination, were driving forces for dam building, their construction and completion often became synonymous with development, economic progress, and even nation building, especially in the recently decolonized countries of the Global South. Dams symbolized not only humans' ability to control natural resources but also the aspirations and optimism of newly created states, particularly in Africa. In South America, the binational Itaipú Dam project was launched by two well-established countries, Brazil and Paraguay; the dam itself was built between 1975 and 1982. It is the pride of Paraguay and provides no less than 90 percent of its energy.

At the apex of global dam building in the 1970s, two or three new dams were commissioned each day on average. The price for this kind of development was high: the World Commission on Dams estimates that between 40 million and 80 million people have been displaced by reservoirs and that the benefits of dams have for the most part been inequitably distributed. Large dam projects have resulted in the loss of forests and wildlife habitats and have diminished aquatic biodiversity. Since the late 1970s, the decline in dam building has been as dramatic as its previous surge, especially in North

America and Europe.[14] However, on the world's largest hydroelectric dam project, the Three Gorges Dam complex on China's Yangzi River, structural work was finished in May 2006, thus providing the stunning antithesis to the deterioration and even breaching of dams in other parts of the world. One historian has aptly called it a "vestige of Soviet-style central planning by specialists who disdain the opinions of affected citizens."[15] The social and environmental costs of the Three Gorges project led to considerable unrest in the 1990s. Today the most important dam building and hydraulic engineering projects are in China and India.

If the current level of public interest in rivers continues, environmental historians will soon have a wider readership. One of the most influential studies to date is Donald Worster's pioneering book *Rivers of Empire* (1985), which focuses on waterways in the American West. Worster argues that the growth of this region—demographically as well as economically—was possible only because of numerous large water projects that dammed and diverted rivers in order to irrigate a landscape that was essentially dry. Ever since John Wesley Powell's *Report on the Lands of the Arid Regions of the United States,* written in the 1870s on behalf of the U.S. government, it has been clear that irrigation was a *conditio sine qua non* for the settlement of the Great Plains and most other parts of the American West.[16] Many scholars and journalists wrote about the importance of rivers west of the Mississippi, but Worster took a structural approach, demonstrating that the need for water not only irrevocably changed many of the landscapes and ecosystems of the West but also led to a redistribution of power and to the rise of new bureaucratic and economic elites. Worster's neo-Marxist approach was inspired by Karl Wittfogel, a German-American scholar of Chinese civilization and architecture who in 1949 offered an ecological interpretation of ancient "irrigation societies."[17] Wherever dams and canal networks were built in the ancient world, Wittfogel argued, a new—and in extreme cases a despotic—elite of bureaucrats came into power and took control of both rivers and people. For such hydrological-political systems and with Imperial China in mind, Wittfogel coined the term *oriental despotism*. Following Wittfogel, Worster saw in the hydraulic apparatus of the American West—the hundreds of dams built throughout the twentieth century, particularly during the 1930s—an industrial variant of the water-controlling societies of the ancient world. The control of rivers transformed not just waterways but society as well and turned the arid West into the "hydraulic West," a concept that is not without its critics.[18]

If Worster's interpretation emphasizes technological control and social transformation, another narrative has also emerged, one that focuses almost exclusively on the (ecological) fate of the river. Even before Bill McKibben's powerful lament *The End of Nature* appeared in the late 1980s, a number of scholars and journalists were writing about "silenced," "raped," or "exterminated" rivers.[19] For these writers, who accepted a master narrative of environmental history as a record of decline, human engagement with rivers inevitably led to despoliation. The Swedish environmental historian Eva Jakobsson observes that such conceptions do not allow historians to fully grasp the complexity of the human-riverine interaction.[20] Even so, in much environmental writing the personalities of rivers have fallen prey to the universalizing forces of modern societies. For Philip L. Fradkin, for example, the Colorado River is "a river no more." He describes the Lower Colorado as "the turgid product of pesticide- and saline-laced return flows from the agricultural fields of Mexicali."[21] Similarly, Blaine Harden's book about the Columbia River is called *A River Lost* (1996), and she refers in its subtitle to the river's "life and death." More recently Ellen E. Wohl has used the term *virtual rivers* to describe streams that have "the appearance of natural rivers but . . . [which have] lost much of a natural river's ecosystem functions."[22] Rivers have thus become for these authors sites of loss and indicators of unwarranted human intervention in an otherwise stable natural environment.

Increasingly, however, environmental historians are beginning to shy away from such reductive oppositions. Instead they have begun to understand humans and nature, technology and the environment, as a continuum. Both river systems and human societies are dynamic forces rather than static entities clashing with one another. In his thought-provoking study *The Organic Machine*, Richard White distances himself from interpretations that identify engineering and management of a river with its "extermination" or loss. "The river," he explains, "is not gone": "We have not killed the river. . . . Nor have we raped the river."[23] White argues that these metaphors and juxtapositions, popular though they may be, contribute little to an understanding of how humans have actually altered rivers and how rivers, in turn, have affected human livelihoods: "We can't treat the river as if it is simply nature and all the dams, hatcheries, channels, pumps, cities, ranches, and pulp mills are ugly and unnecessary blotches on a still-coherent natural system."[24] White stresses that there is no clear line of demarcation between nature and civilization. Twentieth-century rivers are human creations, he asserts, but also have lives of their own that exist "beyond our control."[25]

Mark Cioc takes the concept of a river's life one step further. He calls his history of the Rhine an "eco-biography" and points out that the idea of the river possessing a life or a personality is "not altogether out of step with scientific or commonsense notions of rivers."[26] Rivers, according to Cioc, "seem alive to us"; they even have "a kind of 'metabolism.'" Like White, Cioc insists that the modern multipurpose river is developed but not dead, a word Cioc reserves for streams that can no longer support fish and other types of flora and fauna.[27]

Most historians now discuss rivers in terms of permanent or dialectical interchanges between the dynamics of nature and human intervention. Ideas about rivers and water projects—cultural and technological constructions—have changed both the appearance and the function of rivers over the centuries. At the same time, rivers are themselves agents, providers of energy and resources, and a driving force in history.

OVER THE past few decades more historical studies have been written about American rivers than about all other rivers in the world combined. Part of the reason is the important role that water has played in the history of the American West. Since drought is the rule in many parts of the United States, controversies over river dams and reservoirs—water politics and even water wars—have recurred throughout the twentieth century.[28] Historians of European rivers have focused more on the environmental and cultural aspects of rivers and less on water politics than have their American counterparts.[29] Europeanists have also concentrated on the various social, economic, and cultural functions of urban rivers. This volume is one of the first to offer comparative insights into the history of European and North American rivers.[30] As a group, these essays demonstrate not only the many commonalities but also the contrasts between rivers on both sides of the Atlantic. Social and economic needs, ecological values, aesthetic preferences, and national identities have shaped perceptions and designs of rivers in different regions and countries.

It is exactly this wide range of meanings attributed to rivers that David Blackbourn explores. He correlates the cultural and political constructions of rivers in Germany with their material transformation and argues that these two processes are interrelated. Blackbourn's contribution is also a historiographical one, for his chapter, which is neither a triumphant account of nature conquered by the heroic actions of humans nor an elegiac narrative bemoaning the loss of a supposed natural state, helps readers to understand the broader role of rivers in history. Isabelle Backouche traces the Seine's varying role for different classes of Parisians. By the early nineteenth century, she asserts, the river had ceased to be a gathering place for the social elite and had instead become the center of urban activity and national commerce.

Less important as a political capital than as an industrial one, the city of Pittsburgh allowed pollution of its rivers by regional coal and steel producers for much of the nineteenth and twentieth centuries. Timothy Collins, Edward Muller, and Joel Tarr examine the Allegheny and Monongahela rivers, which converge in Pittsburgh and become the Ohio River. These three waterways were crucial to Pittsburgh's growth and economic vitality. In fact, the perception of the Allegheny, Monongahela, and Ohio as simply part of the area's natural-resource network continued well into the twentieth century. Only recently, the authors argue, has Pittsburgh embraced the three rivers on which it had figuratively and literally turned its back for many years.

Dorothy Zeisler-Vralsted takes on the task of comparing two of the world's largest river systems, the Mississippi and the Volga. Despite the sharp contrast between the American and Soviet political systems, the outcomes of engineering projects undertaken on both waterways during the twentieth century were remarkably similar. Zeisler-Vralsted also notes the contrast in political culture, with localized management of the Mississippi and a more centralized decision-making process in the case of the Volga. Fundamentally, however, both processes were driven by ideologies of modernization and development.

Jacky Girel analyzes the interactions between socioeconomic and environmental factors in the reshaping of the Isère River in the Alpine piedmont. Beginning in the eighteenth century, regional administrators responded to fears of marsh fever and flooding by employing university-trained experts to channelize the river. Many locals were opposed to these drainage projects, which disrupted their traditional forms of agriculture. Girel examines the goals and conflicts that have made today's Isère a striking expression of the nineteenth-century alignment of state power with expert knowledge.

Charles Closmann shows how the growing demand for potable water in the burgeoning industrial areas of Yorkshire and the Ruhr resulted in a delicate balance between economic growth and pollution control. In both regions, coal and steel industries dominated the landscape and the economy, yet the responses of local governments varied. Closmann traces a gradual evolution of laws regulating waterways in the Yorkshire valley that reflected Britain's decentralized political tradition. Competition among local institutions over the waterways proved effective in improving water quality. In the case of the Ruhr, the Prussian state and the river cooperatives were the most powerful players in a much more centralized approach to river management. According to Closmann, their tendency to ignore local input manifested a blatant disregard for the local environment and for ecological concerns.

Especially after World War II, rivers in individual nations such as France and Germany were increasingly viewed as "European" rivers. In his study of the Rhine, Thomas Lekan notes the shift in emphasis from nationalist and aesthetic concerns to a wider focus on ecosystem management, pollution control, and habitat restoration. While the Rhine Commission, that river's multinational political institution, predates European unification by several decades, it established a process that enabled the riparian countries to attain a shared ecological vision.

Ute Hasenöhrl examines the tensions between river tourism in Germany and other interests such as industry, energy production, commercial fisheries, and conservation. All of these interests claim to be working toward an ideal river, but conflicts among them have frequently required compromise. Through a study of the Lech River in southern Bavaria, Hasenöhrl delineates these clashes and concludes that aesthetic or ecological objectives have generally been subordinated to the demand for increased hydroelectric power.

The book closes with Steven Hoelscher's analysis of photography and tourism in the Wisconsin River Dells, one of Chicago's recreational hinterlands since the late nineteenth century. He examines the photographs of Henry Hamilton Bennett, showing how Bennett's work in a very real sense created the Dells as a tourist destination. By means of Bennett's photographs, this part of the Wisconsin River—once the site of sawmills and lumber camps—was transformed into a tranquil, picturesque riverscape.

As these chapters demonstrate, waterways have been shaped over time by varying interests, values, and goals. Their constant physical alteration as well as their ever-changing meanings have influenced human history and will continue to do so. By studying the historical changes in and around rivers, historians can also add to the debates now under way in many countries on how rivers ought to be "restored." Their narratives show that restoration itself is a historically fraught category.[31] When restoration is the objective, how can we determine which of the successive stages in a river's existence is the one that engineers and conservationists should seek to recover? Without knowing a river's history, such a question is impossible to answer. And, as the chapters in this volume make clear, simply expecting a waterway to be restored to its "original" state no longer suffices.

It is our hope that historians will find this volume useful both for its findings and for the questions those findings will generate. Future research on rivers in Europe and the Western world may investigate how techniques of river management have circulated among different cultures. Even if rivers took on national meanings, methods of managing those rivers were often transnational.[32] Another theme that emerges from this collection of essays is the degree to which governments have participated in the constant alteration of rivers, in many cases accelerating the pace of riverine changes, displacing local populations, and dismissing local knowledge. Comparisons between individual rivers and the political authorities that have regulated them would thus be likely to produce interesting analyses. One other promising angle, of the many that could be listed here, is the legacy of colonialism and imperialism for rivers worldwide. Were environmental practices part of the hegemony imposed by colonizers on the states or regions they dominated? What effects did unequal power relations and territorial struggles have on waterways and the communities they sustained? Because rivers have had such a powerful hold over mankind and vice versa, historians have at last begun to exercise their own hold on rivers.

2

"Time is a violent torrent"
Constructing and Reconstructing Rivers in Modern German History

DAVID BLACKBOURN

Rivers generate myths and legends. The ancient civilizations that flourished along the Euphrates, Nile, and Ganges associated these waterways with fertility and the divinities that controlled it—divinities that had to be propitiated, lest the blessings of seasonal abundance be withheld. For it is a feature of river spirits that they are willful. As T. S. Eliot wrote in the *Four Quartets,* "I do not know much about gods; but I think that the river / Is a strong brown god—sullen, untamed and intractable."[1] Rivers have also provided a favorite metaphor for history. You cannot step twice in the same river, said Heraclitus. "Time is a violent torrent," wrote Marcus Aurelius.[2] When the nineteenth-century German historian Leopold von Ranke suggested that history flows like a river, it was already a cliché; and so was the tendency of Ranke's contemporaries to compare disorder—revolution and, in the German case, the "Slav menace"—to a "flood" against which dikes had to be erected.[3] Both ideas can be found in a famous passage of Machiavelli's *The Prince,* where history (under the name of "Fortune") is likened "to one of those violent rivers which, when they become enraged, flood the plains, ruin the trees and buildings, lift earth from this part, drop it in another." But, says Machiavelli, "it is not as if men, when times are quiet, could not provide for them with dikes and dams so that when they rise later, either they go by a canal or their impetus is neither so wanton nor so damaging."[4] Machiavelli's political metaphors reflected his own experience with attempts to divert the Arno at Pisa in the early sixteenth century.[5] But if philosophers and even historians were mindful of hydraulic science, the opposite was also true. Certainly in Germany the justification and celebration of schemes to "tame" unruly rivers were often couched in heavily metaphorical language. The language of hydraulic technocracy was anything but neutral, a point that is not lost on the growing body of scholars who take rivers as their subject.

Why are there now so many river historians? One answer, obviously, is that historians are reflecting a broader concern with the environment. We are the academic counterparts of those popular books and Weather Channel programs that present storm and flood stories as a warning about hubris in the face of nature. They turn their raging rivers into celebrities; we settle for writing river biographies. But I think we can pin down this shift in historical interest more precisely than that. Back in the 1960s and 1970s, when "modernization" was still regarded as a good thing, like the high dams that symbolized it, cutting-edge history also operated under the star of modernization theory. Social science history was also going to be remade into a new and better discipline through quantification and large-scale projects. Think of Charles Tilly's *Big Structures, Large Processes, Huge Comparisons*.[6] Over the past thirty years, the shortcomings of that kind of history—like the problems of large dams—have become increasingly apparent. It left too much out: human beings, for one thing (especially women), but also a sense of place. Microhistory in its various forms was one response to this, and I see the burgeoning accounts of particular rivers and wetlands as the microhistories of environmental historians, insisting, with the late Clifford Geertz, that we attend to local knowledge.[7] But we must not limit ourselves to local knowledge. Environmental activists in the 1970s urged us to think globally and act locally. That is what environmental historians have done in their work. How could historians of rivers not include a transnational as well as a local perspective? The streams they study flowed across national boundaries, as did all the major rivers in German-speaking Europe: the Rhine, Danube, Oder, and (until 1945) the Vistula. All rivers belong to a larger hydrological cycle, and changes in their flow patterns have long-range effects. The Aswan High Dam deprived the Nile Delta of silt and the Mediterranean of nutrients, with a lasting impact on their ecosystems. Scandinavian dams, by altering the seasonal flow of fresh water into the Baltic, altered the exchange of fresh- and saltwater between the Baltic and the North Seas.[8]

I have already sounded the two themes I explore here: the cultural and political meanings attributed to rivers and the history of their material transformation. Those are the "constructing" and "reconstructing" of my subtitle, processes that were intertwined. I illustrate these themes by looking at three episodes in German history and then conclude by addressing some larger issues associated with writing about rivers in history. Let me start on the North European Plain, geologically a product of the ice sheet that melted some ten thousand years ago. On the lower reaches of the River Oder, east of Berlin, lay an area known as the Oder Marshes. In the middle of the eighteenth century the river moved sluggishly through thick vegetation, creating a labyrinth of side-arms. Twice a year the area was flooded to a depth of ten feet. Home to many species of birds, fish, and animals, it was a place over which mists swirled and where the columns of insects were so thick that they sounded "like a distant drumbeat."[9] The marshes also sustained a scatter of villages built on sandy mounds. The inhabitants of these villages lived primarily by fishing but also produced hay and pastured animals when water levels were low, using animal dung and mud to build protective barriers against the water. Over the centuries, patchwork attempts had been made to tame this "barren, valueless swamp land," "a wilderness of water and marsh."[10] Then came a series of unusually high floods in the early eighteenth century, caused partly by climate change that produced more snowmelt and heavier summer rains and partly by deforestation in the upland catchment basin. A radical plan was now accepted by Frederick the Great, not just to raise the level of existing dikes but to construct a new, twelve-mile channel for the Oder. This channel would accelerate the flow of the river and constitute the first step in a comprehensive draining of the marshes.

Why embark on such a project? Modifying the course of the river was intended to create new land for colonists, to increase the food supply, to counter the effects of floods that had become more frequent in previous decades, and to improve navigability. These projects bore the hallmark of cameralist science, which called for maximizing resources. But inseparable from these aims was the ambition to "tame" nature. "Whoever improves the soil, cultivates land lying waste, and drains swamps is making conquests from barbarism," wrote Frederick.[11] That was the authentic voice of enlightened absolutism, with its desire to order, measure, and discipline—not just soldiers and subjects, land and raw materials, but nature itself. What happened on the banks of the Oder, and at hundreds of other sites across Prussia, was part of a larger project to conquer intractable natural forces. It included campaigns to introduce "scientific forestry," prevent wildfires, anchor shifting sands, and eradicate such animals as the wolf and lynx. But educated contemporaries were especially determined to exert control over watercourses like the Oder Marshes. Rulers and bureaucrats suspected them as refuges not just for wolves but also for army deserters and bandits. Writers on natural history saw dark, disorderly corners of nature where vegetation and animal bodies decayed, emitting noxious and harmful miasmas.[12]

That answers the question of *why* water control projects were undertaken, but there is also a *how* question to be answered. The bold step of cutting such a large channel in the Oder was enabled by new developments in land surveying, statistical analysis, and cartography available to the absolutist ruler. It also reflected advances in hydraulic theory and the mechanical arts. Frederick was advised by Leonhard Euler, a leading mathematician of the day (the Isaac Newton of his generation, according to a later panegyrist). Euler was also a lifelong friend of Daniel Bernoulli, who in 1738 enunciated Bernoulli's Principle, which states that the velocity of water forced into a narrower channel will increase.[13] Both men were Swiss. But it is no accident that the main architect of the project, Simon Leonhard Haerlem, was Dutch, for the Dutch were Europe's acknowledged experts in the engineering of waterways. It took seven years to dig the new channel by human muscle power. The enterprise, opposed by aristocratic vested interests before it began and dogged by financial problems, was also hampered in its execution by outbreaks of malarial fever, high water that swept away half-finished dikes, and local resistance. These problems were overcome only by drafting in soldiers as laborers, guards, and overseers. In July 1753 the Oder flowed for the first time through its new bed. After a group of dignitaries from Berlin had traveled through the "New Oder," Haerlem and his collaborator, Captain Petri, reported triumphantly, "There is really no doubt that all the former enemies and detractors of this work, which transcends their own horizons, should be ashamed of themselves."[14] Frederick himself, beholding what had been done, boasted that he had "conquered a province in peace."[15]

Even as the river was reconstructed, a legend was constructed: a heroic narrative of conquest. A century later the dike inspector Carl Heuer wrote some verse describing how a "mighty vassal" had "stormed through house and home" until a royal hero "drove him from the field." This "enemy" was the River Oder; the hero was Frederick the Great, who had "cast him in chains."[16] The new settlements on the reclaimed land—the brave new world of dikes, ditches, windmills, manicured fields, and meadows—were a part of this epic story. Later commentators, looking down on the former marshes from the surrounding heights as Frederick had done, always painted the same picture. This was a "blooming province," a "green land in the sandy marshes."[17] And, indeed, remaking the river did create fertile land for raising livestock and cash crops. The former marshes provided almost laboratory conditions for "improved" farming on the English model. The German pioneer of scientific agriculture, Daniel Albrecht Thaer, settled there to write his *Principles of Rational Agriculture.*

The reclaimed Oderbruch not only supplied food to Berlin; as a by-product of livestock raising, malaria also disappeared from the area. But so did a rich wetlands habitat, together with its former inhabitants. There is no reason to idealize the local people's hard lives, which were constrained by both feudal obligations and market forces. But residents of the floodplain were not so irrational as officials assumed. Their economy was geared to the normal cycle of twice-yearly floods; they had evolved small-scale local solutions that permitted them to fashion a livelihood from the waters until large-scale "improvement" came along. What happened on the Lower Oder corresponds closely to the process described by James C. Scott in *Seeing Like a State,* a critique of technological hubris and state power riding roughshod over local knowledge.[18] Meanwhile, the first generations of colonists struggled to turn abstract plans into actual settlements. Disease and heavy labor culled the settlers' ranks; infections killed their animals and destroyed their crops. These events and experiences could be and were accommodated within a powerful narrative of pioneer fortitude. We should remember that the New World was much in the minds of these settlers: across the river, on the lower reaches of the Warthe, similar colonies on reclaimed land were given names like Florida, Charlestown, and Saratoga.[19] Environmental historians, of course, see the problems that arise when humans migrate across a continent with their livestock and seed corn, carrying pathogens with them as stowaways. But the biggest problem was continued flooding in the 1750s, 1770s, 1830s, 1880s, and beyond, the unintended consequences of the improvers' work. In Ernst Breitkreutz's formulation, the conquered river still "rattled its chains mightily."[20] Each time flooding recurred, engineers and officials proposed what they believed was a permanent solution to the problem. And each time it was presented as a solution that would finally surmount the ignorance, or engineering mistakes, or political constraints of previous remedies, down to Werner Michalsky's claim in 1983 that, through East German planning, "the centuries-old dream of humanity to control the forces of nature ha[d] been realized under socialist conditions."[21] In reality none of these supposedly definitive measures—not raising the dikes after the inundations of the 1770s, not blocking off the "old Oder" after the 1830 flood, not the river "correction" in the 1850s, not the advent of steam pumps and dredgers, not the repeated changes in dike associations—*none* of these prevented floods, which were now a threat to the work cycle rather than an integral part of it. Over the course of two and a half centuries no security against the water could be established. Instead, in a sequence familiar elsewhere (along the Mississippi, for example), floods on the Oder were no longer annual, but when they did occur (e.g., in 1897), they were catastrophic.

At Letschin near the Oder stands a monument to Frederick the Great's achievement. At Karlsruhe on the Rhine there is a similar monument to "the man who tamed the wild Rhine." The man in question was a Badenese engineer and army officer, Johann Tulla. The "wild Rhine" denoted the river that emerged from Switzerland, then turned north and flowed for 180 miles through the great rift valley of the Upper Rhine plain. It moved first through countless shallow channels between gravel banks and wooded islands—the Rhine painted by Peter Birmann at the beginning of the nineteenth century, more like a series of lagoons. Farther downstream the river wound across the floodplain in great curls and loops, carving out a new main channel at high water and drowning land, even whole villages, in the process. This problem became more acute in the 1700s, a result once again of both climate change (the "Little Ice Age") and human actions.[22] There was increased settlement on the plain, as woods gave way to farmland. Under these conditions, local efforts to regulate and dike the river, which began in the fourteenth century, only increased the force of the high water and transferred the problem to the next village downstream. A kind of hydrological leapfrog was at work, forcing many villages to move to higher ground in the late eighteenth century.[23]

Enter Tulla, born in 1770, recognized as a precocious talent, and sent at state expense on a European grand tour of groyne dams and bucket wheels—a muddy tour. Three months into his trip he requested more money "in order that I may replace my coat and other necessary articles of clothing which I have ruined unbelievably."[24] Tulla benefited from a growing technical literature and belonged to a larger, more self-confident generation of river engineers. It was while serving his apprenticeship on a Swiss river-correction project in 1809 that Tulla first imagined a wholesale "rectification" of the Rhine. Three years later a memorandum proposed that the river be "directed into a single bed with gentle curves adapted to nature or . . . where it is practicable, a straight line."[25] A faster flow of water would cause the river to cut a deeper bed, preventing flooding and making agriculture more secure. An essential precondition was created by French armies. In simplifying the map of Germany, as Tulla was to simplify the map of the Rhine, French armies provided both motive—an additional motive—and opportunity. The destruction of the Holy Roman Empire made Baden four times larger and set it on a path of state building aimed at integrating new territories along its major axis of communication: the river. The disappearance of numerous tiny principalities also eased the treaty making between German states, and with France on the left bank, that was necessary before such a project could be undertaken.[26] The possibility of fixing the border was an added inducement. As a French engineer from Strasbourg put it in 1814, "Everybody agrees that all boundaries should be as fixed and invariable as possible; yet what is more variable than the middle of the Rhine . . . ? [It] changes its course every year, sometimes two or three times. With the floods, an island which in the spring was French, is German the following winter, then becomes French again in two or three years."[27]

Work on the Rhine "rectification" started in 1817. The Rhine between Basel and Worms was shortened by fifty miles, a quarter of its length. Dozens of new channels were cut, and more than twenty-two hundred islands were eliminated. This rectification of the river, begun in the wake of the French Revolution, was not completed until after Germany was unified in 1871. As on the Oder, the early stages of rectification met with both political opposition and popular resistance that required troop deployments. Tulla himself died in 1828, long before the project was completed, but not without disparaging his critics as "people of limited views," "ignorant and sometimes malicious people," and "people who are not specialists."[28] The Rhine would be made and remade again, but Tulla had begun the process that turned it into the familiar modern artery, for better and worse. The lower water table changed the climate and usage of the floodplain. Farmers displaced reed cutters and the fowlers who had worked rich birding grounds. These survived only as place-names such as Entenfang, the busy streetcar interchange between Karlsruhe and the river that was once a place alive with ducks. The gold extraction that had gone on since Celtic times also vanished in the wake of the project, leaving only place-names such as Goldgrund and cultural traces like Wagner's *Rheingold*, first performed in 1869 just as the gold extraction itself disappeared. Construction and new hydrological conditions not only disturbed the gold-bearing gravel banks but also destroyed the breeding grounds and resting places of fish, especially migratory species such as lamprey, sturgeon, and salmon. Their sharp decline, which was accompanied by the near total loss of the riparian wetland forests, stands as the symbol of an ecologically impoverished, homogenized river.[29]

The transformed Rhine conferred real benefits, but the price was high. Not all the negative consequences can be attributed to Tulla. The evidence of fauna and flora in the river and on its banks strongly indicates that the degradation intensified after later changes: further Rhine canalization and dredging that permitted larger vessels to navigate the river and led to the heavy boat traffic we see today; industrial development on the riverbank, which brought pollution and sudden changes in water temperature; and hydroelectric plants that impeded migrating fish.[30] These changes were not a part of Tulla's vision, far-sighted though it was for its time. Nor was it Tulla who closed off the old side-arms of the river. They had provided retention basins for floodwaters, but later engineers who eliminated them effectively increased the probability of serious flooding on the Lower Rhine. Yet in at least one case a dramatic change can be directly traced to Tulla. Along the southern stretches of the Upper Rhine, Tulla's plan succeeded too well: the river not only ran deeper but also scoured its bed, in some places deepening it by more than twenty feet. This scouring not only turned wetland into dry grassland of buckthorn and bramble but also created new obstacles to navigation. Further corrections were partly a response to these unintended consequences.[31] In other ways, too, Tulla's successors upheld the logic of his grand design. He made a world that was safe for cultivating sugar beets; the sugar refineries followed. Tulla promised security for dwellers on the Upper Rhine floodplain; subsequent generations redeemed that promise on a large scale but at the cost of ecological damage and greater insecurity downstream.

The effect created by the straitjacketed river had its admirers. In the nineteenth century August Becker found the plain of the Upper Rhine "so fruitful and luxuriantly green, so thoroughly planted and cultivated, that it seems like one great garden."[32] Not all cultural constructions placed on the new Rhine were so favorable. In the late eighteenth century Goethe had fished from one of the former islands in the river, describing how he "brought the cool inhabitants of the Rhine mercilessly into the pot to fry in the sizzling fat."[33] His nineteenth-century successors caught views and vistas rather than fish—but not on the Upper Rhine. It was the Middle Rhine between Bingen and Koblenz that drew visitors to admire its rocky gorges and romantic legends, for this was where (in the revealing phrase of one contemporary) "the Rhine begins to look like itself."[34] The guidebooks characterized the river farther upstream, Tulla's Rhine, as dull; it was not the "real" Rhine. So most tourists left the Rhine at Mainz and headed inland. It sounds like a paradox that the cultural "construction" of the wild, romantic Middle Rhine occurred simultaneously with the taming of the "wild Rhine" elsewhere. But the paradox is only apparent. For the enjoyment of those sublime and stirring tourist panoramas was itself made possible by channel dredging and the dynamiting of dangerous reefs, which created a waterway safer for steamships and their passengers.

I turn now from the lowland river valleys to the large dam-building projects, begun a century ago, that transformed the upland streams. Dams already existed in mining areas such as the Erzgebirge during the early modern period, but Germany came late to the modern era of dam building. Its first monument was the Eschbach Dam near Remscheid, completed in 1891. Although extremely modest by later standards (about 1 million cubic meters of water impounded), Eschbach initiated a "golden decade" of German dam building in the 1890s.[35] Its architect was Otto Intze, professor and civil engineer in Aachen. Born in sandy Mecklenburg, far from the green and rainy valleys of the west where he made his name, Intze is usually seen as both the "trailblazer" and "grand master" of German dams. Richard Hennig included him in his book on famous engineers, alongside Marconi and the Wright brothers.[36] This recognition was not for any major formal innovation. Like Tulla, Intze traveled to observe what others were doing. He was impressed by the masonry gravity dams he saw in France and adapted their elegant contours to dams in Germany, adding a few signature features of his own (such as the "Intze wedge"). Intze was notable above all for his tireless advocacy, at a time when foreign dam breaches had caused serious concern, and for the energy with which he seized opportunities to showcase these modern marvels, first in the small river valleys that fed the Ruhr and Wupper, then in the Eifel and the Silesian highlands. When he died in 1904, twelve Intze dams were in operation, ten were under construction, and twenty-four more were later built from his blueprints. Intze had trained the engineers who took Germany belatedly into the era of reinforced concrete dams in the 1920s, and he or his pupils had designed two of the three largest German dams built before 1914, in the Urft, Möhne, and Eder valleys. Celebrated during their construction and opened to great fanfare, these dams were invariably described as "giant," "colossal," and "mighty."[37] Like ocean liners and Zeppelin airships, they were a part of what Hans Dominik in 1922 called the "Wonderland of Technology," and the sense of awe they inspired in journalists, popular writers, and large numbers of visitors is indisputable. When it comes to dams, David Nye's argument about the "technological sublime" as a fruit of American exceptionalism is simply wrong.[38]

All dams regulate the seasonally uneven flow of water, but they do so for widely different purposes. One of the most common purposes, irrigation, played almost no role in Germany, as the agrarian lobby endlessly complained. It is true that German agrarians turned endless complaining into a political way of life, but here they had something to complain about.[39] The first German dams were commissioned, paradoxically, by the owners of small industrial concerns who harnessed simple water power in upland valleys. They hoped, by securing a more regular head of water, to compete more effectively with the coal-fired industrial giants of the Ruhr. But that quickly turned out to be a vain hope because the titans of industry cornered the market on water, too. Besides, the future would not rely on waterwheels but on turbines that would generate hydroelectric power. This hydroelectric strategy also had anti-coal connotations. Hydroelectric power was, in the words of one advocate, "a powerful, continuous form of energy independent of strikes, coal syndicates and petroleum rings."[40] In fact there was a utopian strain to German enthusiasm for hydro, or "white coal." It was hailed as cheap, clean, and modern, a renewable and "progressive" form of energy that came from the liberal German south and would emancipate the "little man." By the late 1920s, however, southern hydro had been fully integrated into northern electricity combines headquartered in the Ruhr or Berlin, and extravagant social claims had proved illusory.

Dams were also built to provide drinking water for growing cities, to replenish water rapidly disappearing from rivers like the Ruhr, to aid navigation, and to hold back floodwaters. New expectations drove demand, but so did the need to offset the effects of earlier interventions in the hydrological cycle. Drinking water was in short supply because of local pollution and falling water tables, a result of mining and overpumping. Sharply rising industrial demand for water led to the relentless dam-building program of the Ruhr Valley Reservoirs Association. Dams designed to maintain minimum water levels in summer—the holy grail of inland shipping interests—were needed because of river regulation and canal building, on which German governments spent DM 1.5 billion between 1890 and 1918.[41] Reservoirs that raised minimum water levels were sometimes necessary simply to counteract the effects of earlier measures that had caused rivers to scour their beds. The classic instance of using today's new technology to undo the malign effects of yesterday's new technology was building dams to retain floodwaters. This was a special enthusiasm of Intze's (he gave private lectures on the subject to Kaiser Wilhelm II after the floods of 1897), and it was the aim of the dams he designed in Silesia. While laying the foundation stone for one of them, he invoked a familiar metaphor: "It is necessary when dealing with rivers that carry large masses of water . . . to present the water with a battleground so chosen that humans come out the victors. This battleground against the forces of nature should be the creation of large reservoirs."[42]

But were floods "forces of nature"? Their proximate cause was certainly a combination of topography, hydrology, and meteorology. But floods also resulted from deforestation and the regulation of mountain streams, and the damage they caused downstream was intensified by human settlement patterns. A surprising number of contemporary experts recognized this. And because turning the clock back seemed impossible, they offered dams-plus-reforestation as the best solution to previous mistakes. In 1907 a Professor Nussbaum demanded a state program to dam every German river valley that could be dammed (his budget calculations ran to the year 2012). This was a "duty," he argued, because deforestation and river regulation had created a "deplorable situation."[43] By the 1930s the failure of dams to provide any substantial protection against floods was so widely recognized that Intze's disciples began to suggest that even he had harbored doubts. If so, he had an odd way of showing it. Dam building was littered with false hopes and exaggerated claims—as one critic put it, engineers were "extraordinarily generous with promises about tomorrow."[44] The only certainty was that future water demand would exceed current estimates, so that new and bigger dams were projected even as current ones were being built.

The largest German dams were small by world standards. Hoover Dam, completed in 1936, compounded almost two hundred times as much water as the Bleiloch dam, completed just four years earlier.[45] Because the German dams were smaller, and because the climate and usage patterns also differed from those of the United States, German dams did not lead to familiar twentieth-century woes such as salinized irrigation regimes and related problems or high levels of silting and evaporation. Valleys and villages were drowned, but the numbers of people forced to relocate even by large dams were in the hundreds, not the hundreds of thousands. This is not to minimize the painful, slow-motion disappearance of their homes, a spectacle that captured contemporary imaginations as it does our own. (Interestingly, we find no cases of physical resistance, only resignation and long legal struggles over compensation.) Nor has there been in Germany, as in some neighboring countries, any recorded instance of reservoir-induced seismic activity. Many German dams were destroyed by wartime bombing, with heavy loss of life (especially among Russian slave laborers), but to date no German dam has failed, thanks to a combination of conservative design, careful construction, rigorous inspection, and good fortune. That stands in stark contrast to the collapse of two hundred dams worldwide during the twentieth century, thirty-three in the United States alone between 1918 and 1953—events reported by German engineers with a mixture of pained incredulity and schadenfreude.[46] Still, some German dams are less secure than has long been believed, especially Intze's, and many are—like former nuclear power plants—a huge financial encumbrance.

If their negative consequences pale by comparison with those of the Aswan or the Volta, German dams still brought about major disturbances in riparian morphology and ecological structures, with effects on entire river basins. They altered the dynamics of water flow and sediment deposition; all dams silt up, the only question is how fast. Relieved of the heavy load of silt, rivers rushed with greater force through their beds, scouring them out and thus requiring a further intervention. And while reservoirs tempered the dehydrating effects of drainage and regulation, tunneling through rock in large dam projects had the unintended consequence of siphoning water from land overhead, which lowered the water table. At the dam site, where the reservoir basin was scraped, dynamited, drilled, and sealed, new bodies of water were created that bore only superficial resemblances to natural lakes, as the pioneering limnologist August Thienemann noted in 1911.[47] Frequently changing water levels produced a terracing effect on the shoreline, eliminating a riparian zone where flora and fauna could establish themselves. Reservoirs were more vulnerable to eutrophication. Fish migration was disrupted, and species were increasingly homogenized until they were fully managed under programs of "bio-manipulation" partly designed to prevent eutrophication. It is true that reservoirs also attracted new water-dwelling insects and especially migratory birds, but only those who expect human interventions to produce nothing but environmental negatives would be surprised by that. After all, the Salton Sea—product of an American hydrological scheme gone horribly wrong—now attracts more bird species than any other site in the continental United States.[48] Today the ecological management of German reservoir areas has become as unquestioned as the "green" retrofitting of the dams themselves.

Ironically, it was dam advocates who first suggested, as a sop to critics, that reservoir surrounds might serve the cause of nature conservation. By the 1930s, this had become a commonplace idea. What nature conservationists failed to do for most of the twentieth century was to stop or even delay dam building. While this failure was in part a matter of powerlessness, it also reflected an essentially aesthetic concept of landscape preservation. Hard-nosed engineers could live with that. One of them defended the Walchensee hydro project in Bavaria by conceding that some projects impinged on "majestic wilderness" but then arguing that "sacrilege against the precious gift of natural beauty" was not inevitable; things could be managed so that "the scenic landscape is preserved wherever possible."[49] A consensus emerged. It suggested that dams were acceptable if they merged harmoniously into their surroundings, thus shifting the question from "should we build?" to "how should we build?" The focus on landscape aesthetics opened the door to another question: What if dams not only spared natural beauty but enhanced it? Those who maintained that these man-made structures improved on nature pointed to the harmonious elegance of the retaining walls and especially to the reservoirs behind them. Where critics saw a crude horizontal line that dominated the eye, enthusiasts saw a "majestic" body of water that allowed for the play of light and mirrored the surrounding hills.[50] This sentiment was widespread. "Reservoir romanticism" brought together those who celebrated technology and nature conservationists who believed that the new structures enhanced an authentically "German" landscape.[51]

WHY SHOULD historians be interested in rivers? Let me suggest three reasons. First, rivers are repositories of myth and legend. They have been endowed with spiritual significance and often incorporated into the way that human communities—whether small groups or nation-states—have perceived themselves. I have tried to show how Germans constructed a new set of meanings for their rivers as they remade them: the fruitful "green land" reclaimed along the Oder, the "great garden" that emerged after Johann Tulla had "tamed" the Upper Rhine, and the "giant dam" whose shimmering waters were a part of reservoir romanticism. What do these ways of describing human handiwork have in common? Hubris, of course—the language of conquest and mastery runs through them. But it is hubris with a particular twist, for these descriptions suggest that Germans could somehow improve on nature while preserving it. German nature, the putative cradle of national character, was the cultivated landscape of meadows, fields, and domesticated streams, the "great green garden of Germany" as August Trinius called it in 1916.[52] Through the period of National Socialism, the word *green* often denoted a sense of German superiority. And the farther east one looked, to the Vistula and beyond, the more strident were the claims that Germans (unlike, say, Slavic peoples) could tame rivers and simultaneously create a "green garden" true to nature's intentions.

River history is valuable, second, because there are few better ways to study the evolution and development of human societies than to look at what they have done with, and to, their waterways. Rivers combine an extraordinary number of properties. They supply water for drinking, washing, and bathing. They irrigate crops and provide calorific energy directly in the form of fish. They are a means of transportation (a river is a road that moves, said Blaise Pascal). They supply water for cooling and other industrial processes. And they drive both simple waterwheels and complex turbines, this being an instance where the wheel really was reinvented. Not all of these many ways of employing the resources and harnessing the energy of the river are mutually compatible. I have looked at a series of rupture points, when German rivers were remade to serve new interests. Three things are worth emphasizing about these moments of transformation. First, knowledge and expertise were crucial, including technological transfer across national borders. Second, in the interplay of interests, large tended to trump small, the central to win out over the local—which is not automatically a judgment on the social utility of these projects: local claims can be parochial, while centralized solutions can represent the general interest. Third, the transformations I have used as examples were coercive: overt, military-backed coercion on the Oder and Rhine, soft power in the case of dams. It is not too much to speak of water wars. To observe how and why German rivers were remade is simultaneously to observe the distribution of power and processes of state building. Those who did the remaking often talked about "mastering" nature. But the human domination of nature also bore eloquent testimony to the nature of human domination—to social power relations.

The third reason to write the history of rivers is because they are a critical element of the hydrosphere. Anyone who believes that history is not limited to the relations of humans with their own species should be interested in what has happened over time to the water that rivers carry, the fauna they sustain, and the riparian habitats they create. In a more generous and imaginative historical world that task would not be left to a group designated as "environmental historians." Rivers flow and do their work whether or not people are present. That is to say, they do what we call "flow" and "work": the river has no name for those things, which are human constructs every bit as much as the image of the river "rattling its chains." And our river histories must inevitably be anthropomorphic. With all respect to Donald Worster, it would be impossible to "think like a river" even if we wanted to.[53] That said, the salmon of the Rhine—to name the most celebrated nonhuman species mentioned in this essay—are far from mute witnesses to history: they have evidence to give if historians are willing to elicit and weigh it.

That brings me to a final question, perhaps the hardest. How do we write about the reconstruction of rivers without falling into one of two equally undesirable ways of telling the story? First is the narrative of benign modernization: progressive human mastery is imposed on a refractory nature. That is a much less tempting framework than it used to be; even the historians of technology have found it unsatisfactory. It is evident that great "modernizing" projects had negative consequences, usually unintended (though not always unforeseen). We can even identify the pattern of thinking, the present-minded solipsism that led successive generations of river engineers to proclaim confidently that *this* time they had come up with a definitive solution. The second way of framing the account is much more of a temptation today, especially to environmental historians. It is an inverted version of the first, a narrative of decline and disaster in which rivers are killed, raped, denatured. This interpretation is no more satisfactory than the first. It plays down the human benefits, just as the confident technocrats play down the losses. And it too often neglects unintended consequences that point in another direction and underscore how hard it is to "conquer" nature—the complex new biotopes created along old side-arms of the Rhine, for example, or the reservoirs that became links in the flyways of migratory birds. Especially problematic is the moralistic binary of before and after. Before human intervention, stable, self-equilibrating rivers; after, denatured pseudo-rivers. A history that takes the environment seriously is bound to turn up warnings from the past, but it is likely to be one-dimensional history (and unhelpful in grappling with present-day problems) if it is nothing but a jeremiad. Not everything is or was a downward glide to perdition.

The almost religious sense of a "fall" is palpable in much writing about humans and the natural world: humankind has transgressed, lost its innocence, and been expelled from Eden. No one has stated the problem with this stance more trenchantly than the American environmental historian Richard White writing about the Columbia River: "To call for a return to nature is posturing. It is a religious ritual in which the recantation of our sins and a pledge to sin no more promises to restore purity. Some people believe sins go away. History does not go away."[54] None of the rivers I have been writing about was remotely "natural" before they were corrected, regulated, or dammed. There is no baseline from which to measure a "lost" natural world. What conservationists wanted to conserve was the status quo at a particular point between one set of human interventions and another—yesterday's "progress," after it had acquired a patina of "naturalness." Each of these rivers has been imprinted by centuries of human activity. They have been, and they remain, hybrids: organic machines. Which does not mean, of course, that we cannot register processes of cumulative degradation. Even then, some changes have been more destructive than others. Water-borne pollution that once led to huge fish kills in German rivers has proved reversible, but many cases of habitat destruction and loss of species diversity have indeed proved irreversible, examples of what ecologists call the Humpty-Dumpty effect. The detail matters, questions of degree matter. Paradise lost, like the fertility cults of the Euphrates, is an agreeable human myth. But it is a myth all the same.

3

From Parisian River to National Waterway
The Social Functions of the Seine, 1750–1850

ISABELLE BACKOUCHE

Today's Parisians often lament that their city has turned its back on the Seine, which seems to flow on with an indifference to urban life. The systematic engineering of the river's banks in the nineteenth century, usually seen as the preeminent symbol of this estrangement, was in fact only one step in a much longer process. The stages of the Seine's gradual separation from Paris began in 1750. The dynamics of those stages transformed a riverine space once familiar to Parisians into one estranged from the city. Within a one-hundred-year period, the uses of the Seine were altered by a royal administration eager to transform the river into a national waterway. During that time, Parisians lost access to the banks of a river that came to have little meaning for the city except as a supply route for goods too heavy or too expensive to be transported by road.

My methodology for analyzing the Seine's transformation relied on three principles. First, I examined many heterogeneous sources—for example, statutes and other regulations issued by the provost of merchants and the king; account ledgers and other financial records; petitions and settlements of legal disputes between users of the river; and development studies of the river and its environs.[1] Second, I studied elements that vary in scale—for example, a pier, a port, the downstream or upstream portion of the river, the whole river within the city, or even the role it plays within a particular region.[2] Third, I adopted a chronology specific to the historical object in question—

the year 1750 marked the beginning of a radical change in how the river was used, while after 1850 new challenges and problems arose.

Such a study, attentive to the diversity of agents who helped to determine the river's development, highlights the complexity of the Seine's transformation between these two moments. In fact, this transformation was not the result of deliberate urban planning but the outcome of a two-part process in which changing ideas about how the river might serve not only the city but also the nation inevitably led to structural modifications of the channel, the ports, and the waterfronts. Procedures and regulations for implementing public works also changed in the middle of the eighteenth century: the identity of those who designed engineering projects for the river, their objectives, and the scale of their proposals were all manifestations of that change.

The number and variety of protagonists also help to explain the magnitude and intensity of both contemporary debates about the river and the conflicts they provoked. Records of these conflicts are invaluable because they allow historians to identify and dissect the various decisions that would ultimately transform the river. Among the participants and factors that influenced these decisions were the municipal authorities and the king, who made powerful arguments that had to be taken into account; the architects and engineers whose priorities emphasized the river's new functions in the city; Parisian citizens and other local users, who expressed opinions that often revealed resistance to change; and, finally, the financial means of those promoting the project. These debates and conflicts constitute a body of knowledge, both bureaucratic and technical, to which can be added documents and engineering proposals from a state eager to manage this important resource. The transformation of the Seine offers an interesting case study of a contest between local municipal authorities, anxious to retain their prerogatives in the capital city, and the national government, which eventually won the battle by exploiting the political upheavals of the French Revolution to increase its control over the rebellious inhabitants of a Paris in turmoil.

A New River in the City

In 1750 several groups were competing for access to and use of the Seine. The great variety of users reflected intensive exploitation of the river dating to the Middle Ages, and this exploitation was specific to the Seine in Paris. First, along the Seine there were many permanent installations: laundry boats (more than eighty), bath boats (fifteen), and fisheries. There were also mills located between bridge arches and water pumps on Notre Dame Bridge and

Pont Neuf.[3] Workshops cluttered the banks and the bridges, while small stores occupied the coveted ground floor of the houses bordering the Seine (a majority of Parisian bridges were then lined with five-story houses that divided the river into autonomous basins). Some trades and professions required water in order to function. Such was the case of the dyers on Rue de la Pelleterie, in the northern part of Île de la Cité. Around the Châtelet the tripe stalls also required access to the Seine, where animal guts were washed before being cooked in large pots inside the slaughterhouse—an activity that caused much concern among municipal authorities, who considered it a fire hazard.[4] Most of the water consumed by Parisians and two-thirds of the municipal water supply came from the Seine. The activities associated with meeting that demand made the waterfront a lively space within the city.[5] All the goods reaching the capital were obliged to be sold in Parisian ports.[6] The Seine was thus a gigantic, thriving urban market where Parisians purchased commodities and luxury goods as well as their daily necessities. Each product had its assigned port, and municipal authorities carefully monitored the rotation of boats in order to avoid shortages (figure 3.1).[7] The towing and dismantling of boats also created a new category of specialized workers eager to protect their privileges.[8] Moreover, the Seine was used for transporting travelers: barges commuted downstream between Paris and Rouen and upstream between towns located on the Seine itself (such as Melun) or on its tributaries (such as Auxerre, Sens, and Montereau). Finally, the river was a place for festivities sponsored by the city or the king to celebrate important events. These included fireworks and public ceremonies or contests such as the seasonal water games organized between neighborhoods that bordered the river.

This profusion of usages expressed a kind of paradox: while the Seine split the urban space, it also brought the city's inhabitants together on a daily basis and constituted a space of Parisian identity whose trace is still visible in the city's coat of arms. By 1750 the growth of the Parisian population had begun to increase demands on the Seine, and the corresponding need for greater regulation of traffic and commerce made it difficult to reconcile the interests of a rising number of users. Competition for space on the river and the waterfront led to expansion into areas that until then had been on the periphery of city life. For instance, downstream, the Île des Cygnes hosted wood merchants, boat demolishers, and tripe butchers in succession. The governor in charge of the water pump known as La Samaritaine, near the Pont Neuf, denounced the pollution caused by both the boats transporting charcoal and the tripe butchers, while bridge managers complained about the mills, which

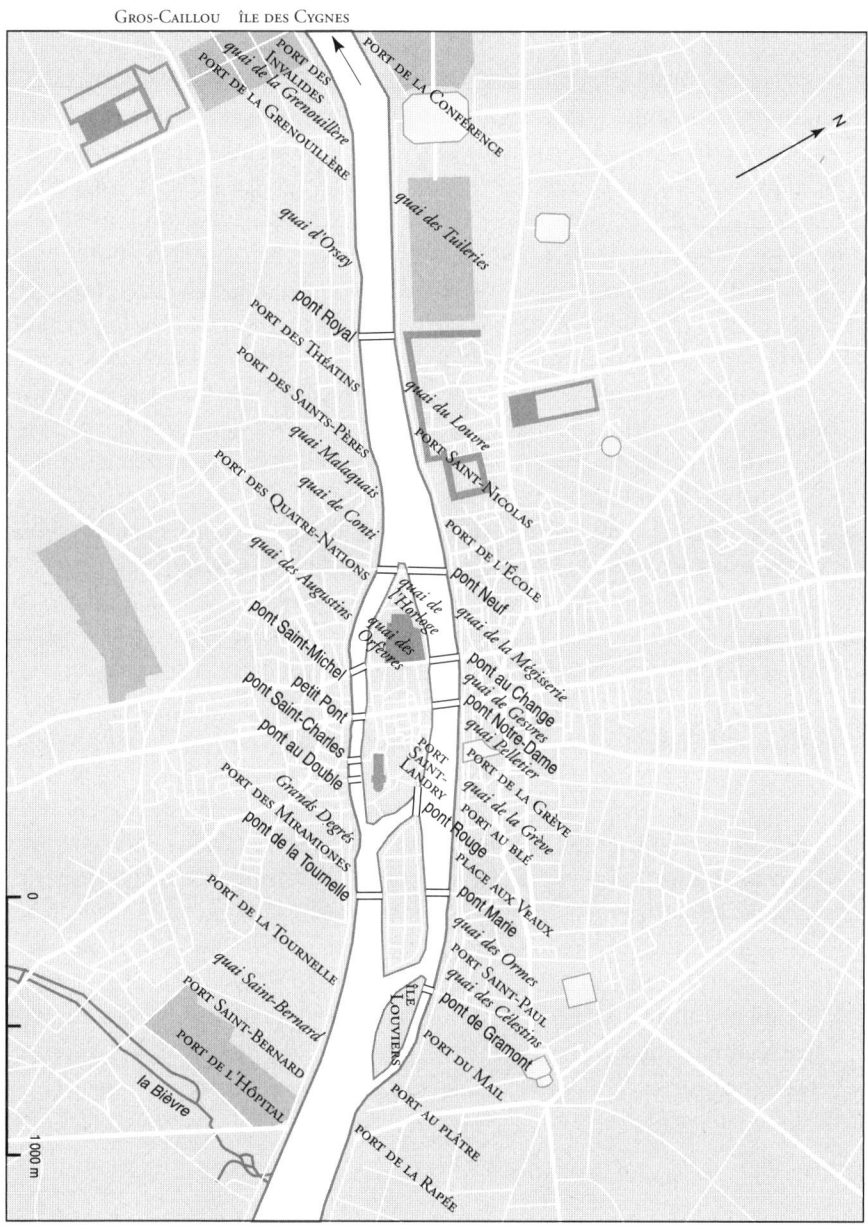

Figure 3.1. Map of Parisian ports and bridges along the Seine. Created by Jacques Bertrand and reproduced by permission of Éditions de l'EHESS.

made shipping and traffic between Pont au Change and Pont Notre-Dame especially hazardous.

The city gradually reduced the number of mills and laundry boats by refusing to renew their leases. Navigation, which was considered the most important activity on the Seine, was given priority over other uses. In the years following 1830, more and more commerce was banned from Parisian ports until the Seine was dedicated exclusively to navigation; all other uses involving local residents had been replaced by a single activity with a national purpose. Once an urban river, the Seine became increasingly alienated from the economic and social life of Paris. Between 1800 and 1820 the development of Paris ports was linked with that of a network of canals in the north of Paris. That network would eventually discourage the populace from traversing the capital city via the Seine in favor of the Bassin de la Villette, another peripheral space that in the nineteenth century would become an industrial and commercial hub (figure 3.2).[9]

The physical shape of the Seine was also being transformed at this time. The great diversity of the intersections between Paris and the Seine—natural banks, high piers, dwellings, and commercial enterprises built along and even over the water—gave way to a series of homogeneous waterfronts designed to improve traffic and prevent damage from periodic, devastating floods. Between 1780 and the early 1800s, houses that lined the bridges were demolished, and bridges were used for traffic only. This destruction did not take into account the communities living on the Seine, which had long contributed to the social vitality of the river and the city. When threatened with eviction, residents of the bridges fought vigorously for their rights, asserting both their ancestral claims and their economic dependence on the river for their livelihoods, and, in the case of the inhabitants of Pont au Change, even going so far as to declare ownership of the bridge.[10] In this respect the city's settlements between opposing parties exemplified the conflict between public and private interests, and the Seine's transformation over the course of the eighteenth century affirmed the preeminence of the common good in urban planning.

An Intellectual Turning Point

The reassessment of the river's traditional uses was also connected to an intellectual evolution that could be seen in the engineering projects that changed the relationship between the Seine and Paris during the second half of the eighteenth century. These projects were developed by individuals from

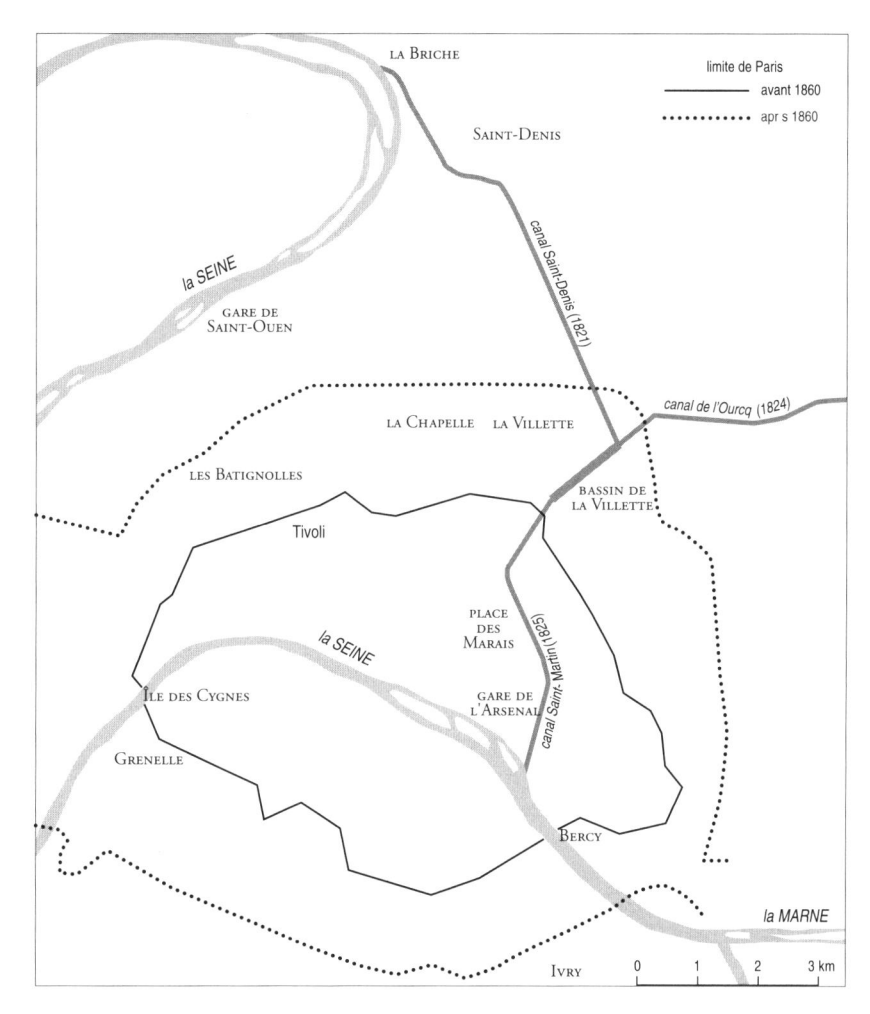

Figure 3.2. Map of the Seine and canals in and around Paris before and after 1860. Created by Jacques Bertrand and reproduced by permission of Éditions de l'EHESS.

various professions; they were lawyers, magistrates, town councilors, doctors, and military officers. While engineers and architects were influential in the river's transformation, they were not the only voices in discussions about how the Seine would change. Consensus was made possible, however, only by having several goals in common: improving navigation in order to make the river a national waterway and alleviating the city's traffic congestion, housing shortages, and related sanitation and public health problems. Despite their heterogeneity in terms of profession, participants in the decision-making process all belonged to the same Parisian elite, and their projects transmitted a homogeneous vision of the city, one that accentuated the opposition between urban and rural environments. Development was guided solely by the specific functions associated with different sections of the city and with the river. The latter, reserved for navigation, would have to be cleared of any obstacle to traffic.

Thus, in the mid-eighteenth century, architects, engineers, doctors, and magistrates alike began to speak of a functionalism that called for the strict adaptation of space to urban activities.[11] This model of urban development came to dominate French city planning for more than two centuries. The Seine was promoted as a site for urban experimentation in which new criteria were superimposed on traditional ideas that up to that point had guided city planning. Within the capital of France, the river had to be an open space that offered beautiful vistas; like any other monument, it had to symbolize an ambitious city and a powerful monarchy; above all, it had to satisfy economic needs without compromising the quality of urban life. This urbanism was the result of an original synthesis and marked the eighteenth century as a turning point: from a capital that served political prestige to a city that met the needs of its 800,000 inhabitants.

The idea of a city whose urban planning combines beautification and development is not unique to Paris. Many French cities of the time undertook large-scale architectural projects in order to make urban spaces more compatible with their practical uses. In Bordeaux, the Château Trompette, which stood on the Garonne, was destroyed and replaced by a wide plaza that overlooks the river.[12] In eighteenth-century Toulouse the attitude toward the river evolved because of a new concern for hygiene. Residents of that city recognized that the influx of fresh air and water brought by the river contributed to the health and equilibrium of the city. It thus appeared essential to take the river into account when planning urban development, and large buildings along its banks were constructed with their windows facing its piers, promenades, and ports.[13] Finally, also during the eighteenth century, the city of Nantes built a modern port in order to control the flow of the Loire and to improve communications between neighborhoods and surrounding towns. As elsewhere, architects were building according to a program that associated both functionality and aesthetics, as shown in such projects as the Quai de Brancas, completed in 1760, and the Île Feydeau.[14]

A few examples of Parisian projects can also help to illustrate the re-envisioning of the urban space and the transformation of the Seine during those years. Changing Paris into a seaport demonstrated the municipal authorities' desire to make the capital a national commercial center. As early as 1765, Claude Siméon Passemant and his associate Bellart, a lawyer, had laid out a project for the Seine that would inspire urban planners for the next two centuries. In order to connect the Seine with the Atlantic Ocean, they proposed construction of diversion canals at each of the ten bridges between Rouen and Paris, as well as swing bridges. The riverbed was deepened so that the channel between Île des Cygnes and Gros-Caillou could accommodate larger boats.[15] These changes intensified traffic on the Seine by eliminating the once mandatory stop at Rouen to divide large cargoes into smaller loads for narrower boats. The Seine then became like the Thames in London; the East India Company even built warehouses in Paris to store merchandise.

In 1762, letters patent announced the construction of two facilities critical to the provision of goods and food to the city: a central grain exchange and a harbor station. The letters emphasized both the beauty and the utility of the planned structures; it was considered important for both buildings "to showcase the capital's splendor" and "to provide Parisian citizens with new amenities and better services"—in other words, to exemplify the principles then being advocated by urban planners. The harbor would be excavated on the Left Bank, near the Hôpital de la Salpêtrière, in order to improve traffic on the river:

> The shelters and natural harbors that have been used in many places along the river, over an area of approximately four leagues, to dock boats bringing supplies of all sorts to Paris have declined; ... bargemen have been forced to tie up loaded boats and to keep them for several months and often for an entire year at many places on the river where they are exposed to all the hazards that can cause a shipwreck, and in this way depriving not only the owner but also the storehouses of Paris; ... those loaded boats as well as those being unloaded cover the entire bed of the river, disturbing the navigation and causing many accidents, sometimes even the loss of other boats and watercraft loaded with wood that continually travel along it in order to maintain the supply.[16]

The harbor would also allow the provost of merchants to exert his control in a single transit space where all merchandise entering the capital could be inspected. This second argument reveals how the control of traffic and sales on the river was changing: agents scattered among the ports were being replaced with a single, centrally located office for the regulation of commodities.

The harbor was therefore financed with taxes imposed on goods entering the city, but the operation turned into an economic disaster, causing Parliament to lead a fierce revolt against the project.[17] In 1765 construction of the harbor was suspended, and two engineers from Paris's famous École Nationale des Ponts et Chaussées, Jean-Rodolphe Perronet and Louis Régemorte, were charged with finding another site for the project.[18] Other objections to the project were raised: excavation of the original site was too difficult (the semicircular harbor had to communicate with the Seine by means of openings at each end, making it necessary for the harbor to be deeper than the Seine's lowest water level in order to remain navigable); the harbor was too far from the city center; and it would eliminate shipping from the heart of Paris.[19] The six merchant corporations of the city took advantage of the interruption in construction to express their disapproval. Angered by the fait accompli, they demanded that users' opinions be taken into account when urban developments were proposed. Their argument was indicative of the diverse issues that urban planners were obliged to manage; the planners had failed to consider that the harbor would be unusable when there was high water or ice, which would prevent boats from reaching their destinations. Moreover, the harbor would cause noxious conditions, which could "be provoked by this sort of lake at the entrance of Paris, whose stagnant waters may, when they are tainted, cause epidemics."[20] Finally, the merchants deemed it unfair to expect the general public to pay an estimated £4 million for a harbor that would benefit only the few individuals who were losing between £3,000 and £4,000 per year.

Encouraged by the general opposition to the project, Parliament issued a judgment on September 4, 1767, that the planned harbor be scaled down (after consideration of the experts' report) and that all agreements with contractors be canceled (with compensation).[21] On February 18, 1768, the king confirmed his original letters patent with new ones that commanded Parliament to register the letters and to obey his orders.[22] This conflict between the king and Parliament not only exposed the latter's desire to limit costs that seemed exorbitant but also underscored the new stakes of urban development and the debates they would undoubtedly arouse. A large project that would affect the entire city—in this case a harbor located on the outskirts of Paris and serving as a central junction for traffic—did not have unanimous support. Instead it emphasized the difficulty of transforming urban space. Though never completed, the harbor station project had a lasting impact, as is evident in several drawings of Paris that include a depiction of the harbor station at the city's entrance.[23] By imagining the harbor as a reality, these maps and blueprints attest to both the significance of the project and the will that once existed to see it realized.

At the Salon of 1789, Charles De Wailly exhibited some blueprints for the transformation of the river. In a proposal titled "Plan général du projet des embellissements de Paris" ("General Overview of Beautification Projects for Paris")—the lengthy subtitle of which is very specific about the real concerns of its designer—De Wailly reunites the three islands of the Seine.[24] The originality of his project derived from its focus on connecting the riverine space with the rest of the city. De Wailly's plan, which envisioned one large island linked to the left and right banks by twelve bridges, showed how reuniting the islands could help to improve communication among the city's neighborhoods. The land reclaimed from the river would be the site of new promenades, and the great avenue that De Wailly designed to join the Palais with the tip of Île Saint-Louis would promote a renewal of the Cité neighborhood, which was then considered "unhealthy and without communication, which day after day keeps it deserted." De Wailly's sense of symmetry led him to propose construction of a square named for Louis XVI opposite the one dedicated to Henri IV; it also prompted him to assign each branch of the river a specific function: the southern branch would serve as the harbor station, while the northern branch would be reserved for "active navigation." De Wailly then suppressed the two bridges near the Hôtel-Dieu, which, in the proposal of his student Bernard Poyet, were to be rebuilt on the Île des Cygnes.[25] Finally, he imagined a multifunctional building that would match the hospital: "Because the wheat exchange is known for being insufficient for the supply of Paris, it is necessary to compensate for that by building along the river right after Isle aux Cygnes, a circus that would be symmetrical to the planned Hôtel-Dieu, and where loaded boats could enter and remain if the warehouses happened to be full. During the summer, this basin could be used for games and swimming lessons. The circus's basement, located at the level of the river, could serve as public baths."[26]

Two years later, the architect Pierre Giraud developed a similar project. In charge of evaluating the national assets, he saw vacant land and conceived a project with sweeping ambition, as the name he gave to it implies.[27] Complex and detailed, Giraud's project took into account commercial navigation as well as the city's water supply. Giraud also located his planned harbor in the southern branch of the Seine, which was transformed into a vast warehouse district equipped with dams and icebreakers. In a very systematic manner, Giraud assigned functions to each feature of his plan. The canal between Île Louviers and Île Saint-Louis became a swimming school; the new alignment of piers was to host "a general boat for laundresses, dyers." All bridges and piers were equipped with towpaths. The logic of space following use was more and more evident, and Giraud's project completely ignored the aesthetic concerns that had characterized urban planning of the preceding era.

De Wailly's and Giraud's designs show that their reimagining of the river went beyond the Seine itself.[28] Architects and engineers tried to apply their solutions to traffic problems to the entire city. Their projects synthesized numerous contemporary proposals to improve navigation, such as constructing a canal that would bypass the city and relocating the Hôtel-Dieu. While essentially utopian in nature, these proposals attempted to reconcile competing claims on the river by eliminating islands and increasing interactions between neighborhoods on both banks.

Successive plans for improving the Bastille site between 1789 and 1798 demonstrate that aesthetically symbolic and ostentatious proposals tended to disappear and be replaced by more economically focused projects. In 1789 and 1790, proposals for the Bastille called for construction of a public square and monuments; after 1791, however, the priority became building a port at the end of the avenue connecting that public square with the Seine. As early as 1789, a project by architects Charles-Louis Corbet and Charles Mangin, centered on Place Royale, identified all the key elements of the site, elements that would be reinterpreted by others.[29] Until 1790 development plans for the Bastille focused on the king and on the erection of a national palace. All projects allowed for access roads to the square and attempted to improve local traffic. Cathala's plan for the Bastille was a compromise between an architecturally symbolic square and a space shaped by economic considerations. It depicted a column celebrating the Revolution; the second draft of the plan included a port.[30] As for the architect J. C. Périer, he showed little interest in the square but sought to make his harbor the designated meeting point for suppliers: "All Parisian neighborhoods will have easy access to this building, which will become a warehouse for all commodities; merchants will build the necessary storehouses in its vicinity. Transportation will be easy and inexpensive. Périer's project will also eliminate the inconvenience of transporting merchandise already in the common warehouse by land for several miles and in unfavorable weather; and boats will not have to bring goods back to the capital amid drifting ice."[31]

The exemption from tolls for products stored in Périer's warehouse would create new opportunities for commerce: "It is suddenly giving Paris the physiognomy of commerce. It is the free port that is recognized by enlightened minds as the most authentic stimulant for commerce."[32]

Two other projects designed by Pierre-François Palloy and by an unidentified engineer prior to 1793 confirmed the river's prominence in the planning of the Bastille project.[33] A third project, designed by Le Clerc, the title of which shows a change in priorities, also emphasized the economic aspects of the future square.[34] In a memorandum attached to his plans, Le Clerc recalled the setbacks that the harbor project began to encounter as early as 1760, and he concluded: "Experience has shown that a commercial harbor could not exist within the walls of Paris." The economic dimension was to be present even in the square itself: "In the center, there is a monument to the glorious Nation; its base is a building with a gallery and a distribution room dedicated to exhibiting products of the national industry."[35] Le Clerc foresaw the taxation of goods occurring not when they entered but when they exited the harbor, which would become a "general store." Enlarging the harbor's mission beyond that of providing the capital with necessities and luxury goods, Le Clerc's project asserted another ambition: making the Seine a national waterway.

In the same spirit, in a memorandum dated 1800, Jean-Pierre Brullée suggested creating "establishments of public utility." The majority of these were linked to the river. Brullée wished to fill in the Île Louviers branch and thus enlarge the area designated for the sale of wood. He envisioned a large port upstream that would use La Salpêtrière, among other structures, as a surplus granary. Under his plans, the port would admit more than fifteen hundred boats and other watercraft loaded with wood; "flour mills and factories capable of efficiently producing a multitude of commercial objects" would also be built. Because of the river, Paris would become a gigantic transit port for manufactured goods (and not only for commodities): "The Seine will facilitate the transportation of all objects made in factories, as well as the shipment of the same objects on connecting rivers that lead to almost every corner of the Republic."[36] Far from fulfilling basic needs alone, the river would serve the national economy.

At the end of the eighteenth century, most projects were turning away from urban planning and were instead promoting a river with a national purpose. Integrating France's new economic and political roles, projects addressed both established expectations and diverse needs. A canal around the city to unblock the Seine, a harbor station to encourage trade, and warehouses and ports to increase the capital's influence would be built in the nineteenth century as the result of new economic and political conditions, such as a port authority. Analysis of these eighteenth-century development plans for the Seine and for Paris indicates that traditional chronological landmarks must be revised. One cannot say that the conception and enunciation of new models of city and regional development began in the nineteenth century. In the case of Paris, Baron Haussmann's radical policies during the Second

Empire reinforced the idea that urban revolution began in the nineteenth century, and Parisian historiography has interpreted this moment as a true rupture. On the contrary, it seems to me that the projects discussed here compel us to reexamine Hausmann's project in the light of eighteenth-century city planning and to accept that the chronology of urban development does not always coincide neatly with that of political history.

A River Estranged from the City

The transformation of the Seine within Paris had several consequences. First, the river quickly lost its relevance both as an urban space and as a gathering place: it ceased fulfilling the daily needs of Parisians, and those who once lived, worked, and did business on its banks disappeared. The decline of the river community was also linked to the suppression of the provost of merchants, whose local powers had acted as a counterbalance to those of the king. Starting in 1800, two prefects managed the river within the capital: the prefect of police, who supervised all activities around the river, and the prefect of the Seine, who was responsible for all construction and repairs.[37] Separating the functions of policing and developing the river was not without consequences for the relation between Paris and the Seine. From the nineteenth century onward, the two types of supervision were unconnected and did not follow the same logic. While the prefect of police attempted to ease traffic by transferring activities to the suburbs (a peripheral space that would be annexed to the capital in 1860), the prefect of the Seine tried to improve navigation in order to transform the river into a national waterway, according to recommendations offered by the Ponts et Chaussées engineers, who were already eager to separate the Seine from the city.[38] An example of this appears in the multiplication of bridges spanning the river, which addressed the need to improve urban traffic by increasing the difficulty of crossing the city by boat. Paradoxically, it also minimized the importance of the river by attenuating the physical division it had always represented.

The river's importance as the source of the city's water supply also diminished. In the middle of the eighteenth century, the Seine supplied half of the water consumed by Parisians, and public works were undertaken to increase this quantity. But even these projects opposed two views of the river: the first one favored the installation of fire pumps on the Seine, while the second one advocated diverting tributaries and building an aqueduct (such as the Yvette Aqueduct). Apart from the technical challenges and the greater expense of rerouting tributary rivers, the debates over how to manage the Seine revealed what was really at stake: municipal authorities wished to prevent the Seine from competing with other supply routes in order to maintain its character, which was so vital to the city. Behind engineering and financial questions was a will to promote the river's potential and to save the Seine from becoming an insignificant space. After 1850, however, acute concerns about water pollution and public health had two consequences: a sewerage system was built to protect the Seine's water from contamination, and rivers such as La Dhuis and La Vanne were diverted to supply the city's growing demand. At the same time, epidemics of cholera and typhoid in London made the issue of water purification more urgent; in 1867 the city of Paris created a "study group on sewerage systems and decontamination."[39]

In the long run, river development over the course of the nineteenth century was responding to new problems. After 1840, competition from the railroad forced the city to improve conditions of navigation (by building tide gates and deepening the riverbed).[40] In addition, among the projects with a strictly "urban" justification, only those meant to protect against floods were realized. Yet despite the improvements made to protect the city from high water, Paris turned out to be far from safe. The flood of 1910 showed the city's vulnerability, and the dikes built upstream in 1950 were capable of containing only a fourth of the water that flooded the city in 1910.[41] Even today, warnings against flooding remain common, and city officials have taken increasingly complex preventive measures to prepare for its inevitable occurrence.[42]

Finally, the progressive abandonment of the river has transformed the Seine into a museum, a waterway conveniently used to discover Paris but unconnected to the daily life of Parisians.[43] In the 1970s the construction of roads along its banks further separated the river from the city, because heavy automobile traffic made walking by the Seine both dangerous and unpleasant. Today, the mayor's office is attempting to "reconquer" the river; in a city heavily polluted by cars, the Seine is recognized as a recreational space capable of ameliorating urban nuisances and improving the quality of urban life.

I wish to conclude by asking whether or not the Seine's relationship with Paris can be considered unique. Proactive British politicians have recently restored the Thames to Londoners, as the opening in 2000 of the Tate Gallery of Modern Art—the third largest museum of modern art in the world, built in a former power plant—demonstrated. Canada's capital bears the name of the river that traverses it: *Ottawa* is the English transliteration of *Outaouais River*—a little-known toponymic fact that is of more than minor interest, because it symbolizes the essential role played by water in the city's development. Today the strategic and economic importance of waterways has been largely replaced by their value as recreational spaces, and the Ottawa River is no exception. Since the end of commercial shipping on the city's Rideau Canal, boating has become increasingly popular with both Canadians and Americans. In winter, Ottawa prides itself on having the biggest ice rink in the world: 6.3 kilometers in length with a width of at least 100 meters. In the space formerly occupied by factories, parks have been laid out, government buildings have been erected, and museums such as the Canadian Museum of Civilization have been built. As in Paris, seeing Ottawa by boat has become a main tourist attraction of the Canadian capital.[44]

Cities elsewhere have also started to repair the damage done by separating rivers from the cities through which they run. In Lyon, trees have been planted, small parks have been created, and a marina has been built at the confluence of the Rhône and Saône rivers. In the American city of St. Louis, the levees have been restored and the adjacent floodplain reclaimed, a flotilla of pleasure crafts crowds the river, and a monument to the western frontier has been built on the banks of the Mississippi. In the same spirit but in a grander style, Québec City has recovered the long-neglected waterfront with an ambitious revitalization of the Old Port. Her upstream neighbor Montréal has rehabilitated the old industrial Lachine Canal, creating a landscaped promenade along the banks upstream from the city. Development of canal networks has beautified and benefited the area around Ranstad in the Netherlands by transforming quarries and peat bogs into lakes, creating green spaces, and increasing tourism.[45] Broadly speaking, then, there are two ways of thinking about the place of rivers within Western urban societies: the first envisions rivers as natural routes for regional transportation and shipping; the second seeks to restore access to local residents.

4

Pittsburgh's Three Rivers
From Industrial Infrastructure to Environmental Asset

TIMOTHY M. COLLINS, EDWARD K. MULLER,
AND JOEL A. TARR

Like most river cities in Europe and America, Pittsburgh has looked upon its three principal waterways—the Allegheny, the Monongahela, and the Ohio—as invaluable natural resources to be used in support of economic development and municipal services. These urban rivers have always been as much a part of Pittsburgh's infrastructure as its highways, railroads, mass transportation lines, or electrical grids. For decades many of the natural features of these river systems were subsumed and in some cases destroyed by human activities. Despite these losses, the riparian ecosystems adapted, survived, and now flourish as part of a new vision for the region's future.

Over the course of more than two centuries, the city has forged very different relationships with its rivers, corresponding with three distinct periods of economic life. During its initial hundred years or so, Pittsburgh depended on the rivers for transporting the trade that was its lifeblood; nevertheless, its attempts to manage them physically were minimal and largely ineffective. The city's headlong embrace of industrialism, commencing around the middle of the nineteenth century, radically transformed the rivers' hydrology and ecology. In order to sustain rapid growth, Pittsburgh elites sought to regulate the rivers' flow, shape their banks, and augment the floodplains. They also used river water for both household and manufacturing consumption. In so doing, they conceived of the rivers in this second period as an integral part of the city's infrastructure. Thus the rivers were harnessed, managed, and rationalized for the smooth functioning of the city and its industrial machine.

Turning the rivers into aqueous infrastructure had negative consequences for their ecosystems as well as for the public's health and safety. A second riverine nature evolved as a result of the altered hydrology and massive amount of pollutants that degraded water quality and greatly diminished the diversity and abundance of flora and fauna. In the first half of the twentieth century, progressive reformers sought to mitigate the conditions most hazardous for humans due to water pollution and flooding. They employed additional engineering and conservation measures in their efforts. The subsequent improvements in water quality gradually renewed interest in the rivers' scenic beauty and recreational potential, especially after 1950.

Federal water quality legislation passed in the 1970s and the massive collapse of the region's traditional industries in the 1980s provided the impetus for Pittsburgh to rethink its connection to the rivers. Impressed by the revitalization of urban waterfronts elsewhere in the United States, civic leaders began to recognize Pittsburgh's rivers as potential sites for profitable scenic and recreational enterprises and thus as part of a new regional development strategy. At the same time, community activists began to advocate restoration of the rivers' damaged ecosystems. Promoting a view of the riverine environment—air, water, riverbanks, and the many organisms they support—as a public commons, this group sought to restore that environment to its preindustrial state. While these commercial and conservation agendas sometimes came into conflict, together they inaugurated a third phase in the relationship between the city and its rivers, which have once again become vital to Pittsburgh's economic future.

The Rivers before Industrialization

Flying into Pittsburgh today, a traveler readily spots from the airplane the city's outstanding physical feature—the convergence of two broad rivers to form a third, even larger river. At this junction the tall buildings of Pittsburgh's downtown rise up from a small peninsula formed by the confluence of the Allegheny and Monongahela rivers. These skyscrapers stand like sentinels over the Ohio River, as it begins its westward journey of nearly a thousand miles to the Mississippi River.[1] Far to the south the sources of the Monongahela lie in the Appalachian Mountains of West Virginia. The Tygart and West Fork rivers come down from the mountains and join at Fairmont, West Virginia, to form the Monongahela, which leisurely flows northward for 128 miles to Pittsburgh. In contrast, the swiftly moving Allegheny River begins in the mountains of north-central Pennsylvania and flows briefly north and westward into New York state before turning southwest for 325 miles through western Pennsylvania for its meeting with the Monongahela. These three rivers and the point of land they create together comprise the physical structure around which the Pittsburgh region developed. Three secondary rivers—the Beaver, Kiskiminetas, and Youghiogheny—as well as dozens of tributary streams, creeks, or runs augment this riverine structure.[2]

Before the invasion of western Pennsylvania by French and English military expeditions, traders, and colonists, these river valleys were occupied at different times by various tribes of Amerindians, including the Delaware, the Shawnee, and several tribes from the Iroquois nation.[3] By the turn of the nineteenth century, settlers (with the assistance of military troops) had largely displaced the native population and taken over the land. The rivers supplied the primary organizational framework for this new settlement, providing a transport corridor in drought-free years, an abundance of fish, and fertile land in the floodplains. Moreover, Pittsburgh's location at the head of the Ohio River presented the young city with an opportunity to become the commercial gateway to the newly opened midwestern frontier. Thus the three rivers were not only central to the city's economic vitality and aspirations but also an integral part of people's everyday lives.[4]

Despite their significant role in supporting Pittsburgh's economy, the rivers presented serious impediments to navigation. Water provided an ideal means of transportation, but the fluctuations in flow, dangerous snags and boulders, treacherous riffles, and shifting sandbars all limited commercial uses. At the urging of politically connected local residents, both the state and federal governments funded modest attempts to clear the rivers of obstacles and deepen the channels. They contracted private companies to dynamite rocks in riffles, remove flotsam and jetsam, and establish small "wing" dams to ensure sufficient depth for transport. These efforts to make the rivers more navigable produced only minimal improvements by the end of the Civil War.[5]

One solution was to create a network of canals, which regulated water flow in separate, often parallel waterways. Among the many miles of canalized waterways that the state constructed, the most important was the Pennsylvania Main Line system built in 1834. It was a combination of inclined plane, railroad, and canal extending from Philadelphia to Pittsburgh. For almost twenty years Pittsburgh's commercial life benefited enormously from this awkward system as it snaked into the city along the northern floodplain of the Allegheny River. The value of Monongahela River traffic, especially coal shipping, persuaded the state of Pennsylvania to commission the Monongahela Navigation Company in the mid-1840s to construct the first dams and locks between

Pittsburgh and the West Virginia state line. This modestly successful slackwater venture established privatized toll navigation.[6]

The rivers fulfilled essential functions for the growing city, such as providing water for households and industry and a sink for municipal sewage, storm water, and commercial and industrial wastes. Pittsburgh residents used the rivers for recreation and annually coped with the cleanup required by spring and fall floods that inundated low-lying neighborhoods. Private companies erected bridges to replace the slower, less reliable ferries as urban development emerged on the banks of the Allegheny and Monongahela across from the city at the Point. As a result of the city's expansion in the first half of the nineteenth century, the wooded banks of all three rivers gave way to wharves, docks, and other infrastructure that aided mercantile sales, commercial navigation, and early industries.[7] Nevertheless, the human impact on the natural ecosystem was negligible compared to what it would be in the coming decades.

The Transformation to Urban Industrial Infrastructure

Pittsburgh's hope of becoming a gateway to westward migration and commerce via its rivers and canals evaporated in 1852 when the Pennsylvania Railroad brought the first rail service to the city. This event was the catalyst for a dramatic long-term shift in how civic leaders perceived the three rivers. Instead of transportation routes, they were increasingly regarded as essential infrastructure for industrial production and urban development, and engineers shaped and managed them accordingly.

Industrialization supplanted the city's river-based mercantilism in the three-quarters of a century following the railroad's arrival. The most intensive development occurred in the region's mines and iron- and steelworks. After 1875 the explosive growth of the mass-production steel industry defined the city's transformation. Industrialists sited their mills on the broad river floodplains nestled within the sweeping river meanders. Railroad tracks ran along the mills' landward side (figure 4.1). Other capital-intensive firms dealing with glass, aluminum, railroad equipment, food processing, and electrical equipment, to name the most prominent, also spread to sites along the three rivers' floodplains, stretching for thirty to forty miles from the Pittsburgh Point. By providing cheap transportation for fuel from the region's coalfields and for other natural resources, such as timber and oil, the rivers were critical to the region's industrialization (figure 4.2).[8]

Figure 4.1. (*left*) Pittsburgh's Mon Wharf (c. 1875).

Figure 4.2. (*below*) Coal barges on the Ohio River (c. 1905).

Although the total number of people and vessels engaged in commercial river traffic shrank steadily during the remainder of the nineteenth century, the actual tonnage of goods shipped along the rivers increased. Bulk commodities, such as oil, sand, gravel, and especially coal, along with some finished iron and steel products, dominated this traffic. After the Civil War, the U.S. Army Corps of Engineers took over management of the rivers from private interests and the state. Reshaping them into stable, efficient infrastructure systems produced significant economic returns, since each additional six inches of channel depth allowed boats and barges to increase their cargo by seventy tons. Under pressure from Ohio Valley business interests, Congress in 1875 authorized the construction of locks and dams for the Ohio River. Measurable improvement in river traffic, attributable to the Davis Island Lock and Dam (figure 4.3), located five miles below the Point, intensified political pressure for the "canalization" of the entire three-river system. By the turn of the century, locks had been constructed up both the Allegheny and Monongahela rivers and down the Ohio as far as Marietta. The Army Corps of Engineers finally completed the Ohio River system to Cairo, Illinois, in 1929. All three rivers now consisted of a series of pools connected by navigation channels whose depth was maintained at between six and nine feet.[9]

Figure 4.3. Lock 2 of the Davis lock-and-dam system on the Ohio River.

Controlled river flow and the concomitant growth of industry along the waterfronts changed the river edges nearly as dramatically as the rivers themselves had been altered. The "pools" permanently raised water levels, drowning mudflats and establishing new riverbanks. Industries constructed wooden and cement bulkheads for shipping terminals and flood management. They built large round mooring cells; sprinkled the banks with cranes, conveyor belts, and loading chutes; and added water-intake and waste-discharge pipes. Some companies raised the height of the banks with extensive fill, much of it slag from the iron and steel mills, and a few even closed back channels to nearby islands. Boat-building firms erected launching facilities, and repair yards sprang up along the banks in several places. Shanty boats, sunken watercraft, and abandoned barges littered the edges of the waterway. Railroads lined the riverfronts as well, and in many places steep hillsides narrowed the floodplains so that only the railroads occupied them.[10]

Both municipalities and industries used prodigious amounts of water. They pumped water for household consumption and commercial production while discharging wastewater contaminated with human and industrial wastes back into the rivers. Writing in 1912, N. S. Sprague, superintendent of the Pittsburgh Bureau of Construction, observed that "rivers are the natural and logical drains and are formed for the purpose of carrying the wastes to the sea," expressing with admirable economy the attitude of most municipal officials and local manufacturers toward Pittsburgh's rivers between 1850 and 1940.[11]

This attitude was codified into administrative law for Pittsburgh in 1923 when the Pennsylvania Sanitary Water Board created a classification system that designated each stream in the state as belonging to one of three categories: clean and pure; polluted but not hazardous to public health; or unsafe for drinking and recreational uses, including fishing. Streams in this last category could continue to be used for the discharge of untreated wastes—that is, as open sewers. Pittsburgh's three rivers fell into this third category.[12]

Pollution flows were of three principal kinds: municipal sewage, industrial wastewater, and mine acid drainage. Cumulatively, they all degraded water quality, but their effects differed somewhat with respect to public health and stream ecology. In terms of public health, municipal sewage had the most devastating impact.

Beginning in the second quarter of the nineteenth century, Pittsburghers began to rely on local rivers for their water supply, although residents of some working-class areas continued to depend on groundwater and pumps into the twentieth century. By 1915 the municipal system included 743 miles of distribution pipes.[13] The provision of running water to households and the widespread adoption of water-using appliances such as sinks, showers, and water closets benefited households in many ways, but these developments also exacerbated the problem of wastewater disposal.

For much of the nineteenth century, Pittsburghers placed household wastes and wastewater in cesspools and privy vaults, not in sewers. They continued this practice even after the availability of running water greatly increased the volume of wastewater. In 1881, for instance, about four thousand of the city's sixty-five hundred water closets were connected to privy vaults and cesspools, with only about fifteen hundred linked to street sewers that the city had begun constructing at midcentury. The Pittsburgh Board of Health noted that overflowing privies were a constant nuisance and presented the city with a major health issue.

Such conditions raised the likelihood of epidemics from infectious disease and highlighted the need for improved sanitation and construction of a sewage system. In the 1880s a fierce debate arose over whether the sewer system should be a separate system carrying only household and some industrial wastes (the plan physicians favored) or a combined system that could accommodate both wastewater and storm water in one pipe (the design preferred by engineers). The engineers triumphed, primarily for reasons of cost, and the city began constructing a combined sewer system, building more than 412 miles of combined sewers between 1889 and 1912. That design decision had large implications for the rivers' water quality in the future.[14]

Like other cities, Pittsburgh discharged its untreated sewage from public sewer outlets (145 of them) directly into neighboring waterways on the theory that running water purified itself, diluting and dispersing the wastes. Simultaneously, upstream communities were also building sewers and discharging their wastes into the same rivers. By 1900, more than 350,000 inhabitants in seventy-five upstream municipalities discharged their untreated sewage into the Allegheny, the river that provided drinking water for most of the city of Pittsburgh's population. Even some of Pittsburgh's own sewers discharged into the river at locations above its water-supply intake pumping stations. The resulting pollution gave Pittsburgh the highest death rate from typhoid fever of the nation's large cities—well over 100 deaths per 100,000 people from 1873 to 1907. By contrast, in 1905 the average death rate for northern cities was 35 per 100,000 persons.[15]

Concerned over the growing typhoid mortality and morbidity rates, several professional groups and the Pittsburgh Ladies Health Protective Association formed a citizens' Joint Commission in the 1890s to study the issue of water pollution. Their report of 1894 indicated that Pittsburgh and Allegheny City water supplies were "not only not up to a proper standard of potable water but . . . actually pernicious," and it recommended filtering the water supply of both cities. In 1896 the mayor established the Pittsburgh Filtration Commission, which made a similar recommendation. After several years of delay caused by political infighting and technical problems, the municipal water department delivered the first filtered water to the city in December 1907. Pittsburgh thus joined cities such as Philadelphia and Cincinnati, which took water from their local rivers and filtered it, while cities such as Boston and New York chose to rely on supplies from a protected watershed. Almost immediately Pittsburgh's typhoid rates began to decline, and by 1912, after chlorination of the water supply, the city's typhoid death rate dropped to the level of the national average for large cities.[16]

Filtration provided one safeguard against polluted water, but many sanitation specialists and public-health physicians believed that a further measure—sewage treatment—was required for effective protection. In 1905, responding to a severe typhoid epidemic, the Pennsylvania General Assembly passed the Purity of Waters Act "to preserve the purity of the waters of the State for the protection of the public health." This act forbade the discharge of any untreated sewage into state waterways by new municipal systems, and although it permitted existing systems to continue the practice, it required cities to secure a permit from the state commissioner of health for any expansion of those systems.

In 1910 Pittsburgh sought a permit that would allow it to extend its sewerage system, thus setting up a confrontation with the state agency. The Pennsylvania Department of Health, headed by Dr. Samuel G. Dixon, a follower of the New Public Health doctrine, then ordered a "comprehensive sewerage plan for the collection and disposal of all of the sewage of the municipality." In addition, the department argued that, in order to ensure efficient treatment, the city should consider changing from a combined system to separate sewer systems. The municipality responded by hiring the renowned engineering firm of Hazen and Whipple to study the issue. The report produced by Allen Hazen and George C. Whipple in 1912 asserted that a new sewage treatment plant would not eliminate the need for users downstream to filter their water supplies, since other communities would continue to discharge raw sewage into the rivers. The method of disposal by dilution, they maintained, would suffice to prevent nuisances, particularly if storage reservoirs were constructed upstream from Pittsburgh to augment flow during periods of low stream volume.[17]

Dixon, the health commissioner, found the Hazen and Whipple report an insufficient response to his original order for a long-range plan for a comprehensive regional sewerage system, but most members of the engineering community agreed with its premises. The issues for them were controlling costs and avoiding nuisance. Dixon countered that streams should not be allowed to "become stinking sewers and culture beds for pathogenic organisms." Given the political context, however, and the city's financial limitations, Dixon had no realistic means by which to enforce his order. In 1913 he capitulated and issued Pittsburgh a temporary discharge permit, thus continuing the use of Pittsburgh's rivers as open sewers.[18]

While demands for sewage treatment had been rising, another major pollutant, mine acid, had escaped regulation. Other industrial pollutants caused problems, but the sulfuric acid discharge from coal mines was the most ecologically damaging water contaminant in western Pennsylvania's industrial history. Mine acid destroyed fish life, altered the flora in and along small streams and major rivers, caused millions of dollars' worth of damages to domestic and industrial water users, and raised the costs of water and sewage treatment.[19]

The economic importance of coal production in southwestern Pennsylvania impeded attempts by government to counter the burden of mine acid drainage. The coal industry contended that no "suitable" method existed for the treatment of mine water. In addition, the courts had granted the industry legal protection in the infamous case of *Pennsylvania Coal Company v. Sanderson and Wife* (1886), which concerned the destruction by mine acid of the water supplies of a farm near Scranton, Pennsylvania. In this case the Pennsylvania Supreme Court maintained that "the right to mine coal is not a nuisance in itself" and that the acidic substances had entered the stream via natural forces that were beyond company control. The justices also noted the economic importance of the coal industry, arguing that "the trifling inconvenience to particular persons must sometimes give way to the necessities of a great community." In 1905, when the Pennsylvania legislature passed the Purity of Waters Act, it specifically exempted "waters pumped or flowing from coal mines," as it did again in 1937 in the Clean Streams Act.[20] In general, therefore, as reflected in the failure to regulate sewage disposal and mine acid, the three rivers and their valleys had become and remained industrial infrastructure, engineered and utilitarian. By the 1920s they had been so altered that few older Pittsburgh residents would have recognized the rivers they had known a half century earlier.

Reforming Industrial Nature

The rapid growth and unbridled development of the Pittsburgh region left in its wake a number of social, political, and, as noted above, environmental problems that many early twentieth-century Pittsburghers believed threatened the city's civic order and long-term economic welfare. As with the nation as a whole, local progressive reformers viewed improving the built and natural environments as a means to uplift the society's moral character and health conditions as well as to create a more efficient and successful business community. Although often emphasizing different aspects of the reform agenda, both voluntary citizens' groups such as the Civic Club of Allegheny County and business organizations such as the Pittsburgh Chamber of Commerce agreed on the importance of addressing problems associated with the rivers and riverfronts. Never challenging the primacy of business prerogatives and private property rights, these reform efforts did represent a modification of the prevailing view of the rivers as industrial infrastructure, which was consistent with progressive reform and the conservation movement then gaining momentum at the national level. Pittsburgh's elites increasingly participated in national networks in which they encountered campaigns to conserve forest resources and watersheds, preserve land for parks, and adopt urban planning.

Following the disastrous flood of 1907 and the threat of another a year later, the Pittsburgh Chamber of Commerce appointed a flood commission, chaired by industrialist H.J. Heinz and consisting of representatives from local government, reform groups, and business as well as individuals with engineering expertise. The Flood Commission of Pittsburgh hired an executive director who was strongly identified with national concerns for planned watershed development. In its massive technical report published four years later, the commission projected both pragmatic local thinking and a vision for regional conservation. Predicting that continued urban industrial development and clear-cut logging in the mountainous watersheds of the Allegheny and Monongahela rivers would exacerbate the frequency and intensity of flooding, the commission recommended constructing a floodwall to protect the downtown area, filling in back channels of islands near the city's center, creating seventeen regional impoundment reservoirs, and establishing forest preserves in the headwater areas.

The Flood Commission persuaded the federal government to purchase land for reforestation in the upper watersheds, mirroring the conservation triumphs of President Theodore Roosevelt and his chief forester, Gifford Pinchot. However, it soon became embroiled in a protracted struggle with Congress and federal agencies. In particular, the Army Corps of Engineers resisted broadening its mandate to manage inland rivers for more than navigational purposes. Only successive years of drought, economic depression, and the ruinous floods of 1936 and 1937 impelled local and federal leaders to prevail on Congress to authorize a comprehensive flood-control program.[21]

At the same time that the Flood Commission tackled its work, the Pittsburgh Civic Commission contracted with landscape architect Frederick Law Olmsted Jr. to prepare a plan for Pittsburgh's roadways. Olmsted's work focused primarily on issues of traffic congestion and future highway development, but in his report in 1910 he also lamented the dilapidated condition of downtown riverfronts and urged the city to rethink their possibilities. While he advocated modernization of the Monongahela Wharf cargo-handling facilities, incorrectly predicting a resurgence of general merchandise commerce on the rivers, he also envisioned the rivers as a potential amenity for their picturesque qualities. Modeling his vision on European cities, he proposed promenades and overlooks above the waterfronts, which would coexist with an expanded roadway to channel traffic rapidly around the periphery of the congested downtown area. Moreover, the Boston planner tapped into an ongoing discussion about reclaiming the land where the three rivers converged in order to create a symbolic park and monument at this historical and geographical inception point. Probably recalling the various plans for the preservation of open spaces and riverscapes in Boston, he also urged that several undeveloped tributary stream valleys be preserved and designated as public parks.[22]

For the next three decades Olmsted's successors in planning for Pittsburgh kept these ideas for the waterfronts alive in one form or another, but the industrial and infrastructural utility of the rivers and riverfront lands precluded their implementation. In 1923, for example, a subcommittee on waterways of the Citizens Committee on the City Plan (an elite volunteer group that put forth a city plan in six parts) summed up the business community's view of the rivers when it reported that "navigation interests have a prior right to the use of those portions of the City's water front which can be advantageously used for water transportation, and that no encroachments should be permitted thereon which will interfere with such activities."[23] Further, the subcommittee recommended development of several rail-to-river terminals on the waterfronts in and around downtown.

While a few decades into the century civic leaders still wished to manage the rivers for flood control and city planners envisioned the redesign of riverfront lands, other reformers and conservationists continued to worry about the disposal of waste in the rivers and the hazards it posed to public health. They refocused their energies on the question of water quality. Although by 1934 the drinking water supplies of 80 percent of the state's population were treated, 85 percent of Pennsylvania waterways suffered from various degrees of degradation attributable to raw sewage and industrial wastes. For instance, only 18 percent of the sewage from a population of 920,000 in the Allegheny River basin was treated, and it was the Allegheny from which the city of Pittsburgh drew most of its water supply. Sewage from Pittsburgh and other communities overwhelmed the oxidation capacity of the streams, creating offensive sights and smells on the rivers. Gross pollution levels in streams threatened public health, and while water treatment had sharply reduced typhoid deaths, the death rates from diarrhea and enteritis remained above the national average. Fish were absent from long-dead stretches of the rivers, and chemical pollution fouled the taste of many drinking water supplies. In addition, mine acid drainage increased the costs of water filtration for Pittsburgh residents.[24]

By the end of the 1930s, public concern over these conditions had united environmental groups and associations of outdoor sports enthusiasts in demanding that state and local authorities initiate conservation efforts. In 1937 the Pennsylvania General Assembly passed the Clean Streams Act, giving the Sanitary Water Board power to issue and enforce waste treatment orders to all municipalities and most industries except for coal mines. In 1944 the Sanitary Water Board announced comprehensive plans to reduce the pollution of Pennsylvania streams, requiring all municipalities to treat their sewage "to a primary degree." The following year the board issued orders to the City of Pittsburgh and 101 other municipalities, as well as to more than 90 Allegheny County industries, to cease discharging untreated wastes into state waterways.

The state authorized the formation of the Allegheny County Sanitary Authority (ALCOSAN) in 1946, but political interests prevented any real progress until October 1, 1958, when ALCOSAN's centralized, activated-sludge treatment plant located on the northern bank of the Ohio River below the Point began treating the wastewater of Pittsburgh and eighty-one county boroughs and towns (figure 4.4). A major step had finally been taken to improve the water quality of Pittsburgh's rivers.[25]

Figure 4.4. Pittsburgh industrial waterfronts (1957).

The Pittsburgh Renaissance and the Rivers

Despite the economic revival during World War II after years of depression, Pittsburgh's leaders feared that peace would bring back the declining industrial prospects the city had experienced since 1920. The desire to retain existing firms and attract new companies, especially ones that diversified the region's industrial base, dominated the postwar civic agenda. In 1945, the city initiated a twenty-five-year redevelopment program, the so-called "Renaissance" program, under the leadership of Richard King Mellon, Mayor David Lawrence, and the newly formed Allegheny Conference on Community Development, a nonprofit organization controlled by the presidents of the region's largest corporations. This public/private partnership addressed issues of economic development, renewal of infrastructure, downtown redevelopment, and environmental pollutants such as smoke. By 1960 the national media was heralding the revitalization of the "Smoky City." Even the three rivers received some attention in the Renaissance program. Besides the construction of sewage treatment facilities and implementation of flood-control measures, the city cleared the railroad yards, warehouses, and last remaining residences from the Point and undertook construction of a state park there, which invited contemplation of both the aesthetic and historical significance of the three rivers' confluence. Moreover, recreational boating began to return to the city's waters.[26]

These steps toward the rehabilitation of the rivers proved to be relatively small ones, however, and the flood-protection program only expanded the type of river management begun in the nineteenth century. Despite Olmsted's dreams, prewar proposals by pragmatic urban planners to use downtown's waterfronts for highways and parking lots were implemented during the Renaissance program. The continued inflows of industrial wastes, mine acid, and sewage from the overflow pipes of the combined sewer system of Pittsburgh and other county municipalities prevented any real ecological recovery of the rivers.[27]

In the 1970s the failure to exploit the rivers' aesthetic and public relations potential for two major developments demonstrated the persistent grip of the industrial mindset on the perceptions of civic leaders. Despite its name and its location across from the historic Point, Three Rivers Stadium, which opened in 1970 and became a nationally recognized sports venue, offered no views of either the rivers or downtown Pittsburgh. Similarly, the new convention center that opened in 1977 was designed facing away from the Allegheny and with the rear of the building and its loading docks oriented toward the river.[28] Even the region's principal conservation organization, which emerged in conjunction with the national movement for the preservation of wilderness lands, abjured the city's rivers. The elite-sponsored Western Pennsylvania Conservancy, increasingly active and influential after World War II, focused mainly on protecting environments either well beyond the city limits or on the metropolitan fringe.[29]

Having achieved success with the Renaissance program's smoke-abatement efforts, new sewage-treatment facility, flood-control project, and extra-urban preservation work, the city proudly assumed a leadership role in the national environmental movement of the 1960s. But a plan for the riverfronts called this leadership into question in 1959, at least as far as the rivers were concerned. Pittsburgh's urban planners noted that the Renaissance program had neglected the conservation of the rivers and riverfronts. They recommended several possible sites for riverfront parks and recreation areas. Nevertheless, they still argued that "most of the flat land adjacent to the rivers must continue to be occupied by the industry and commerce which support the City." At best they hoped to "balance" the diverse interests in the rivers, because "the rivers can provide an additional opportunity for recreation without detracting anything from industry."[30] Thus, despite the dramatic postwar renewal program, municipal leaders persisted in viewing the three rivers as engineered, infrastructural systems for industry and urban development. Conservation measures implemented beginning early in the twentieth century had ameliorated only the worst health consequences of this conception and marginally improved water quality and some riverfront lands.

Reinventing the Rivers as Environmental Infrastructure

The collapse of Pittsburgh's industrial base in the last quarter of the twentieth century forced a reconceptualization of the city's economic and social life. Although at the time it seemed as though civic leaders were slow to rethink the city's relationship to the rivers, in fact they began to develop a new understanding in a little more than a decade. After initially scrambling to attract any sort of job-creating businesses to former industrial sites, civic leaders, some developers, and community activists began to envision the rivers as settings for scenic, recreational, and environmental amenities that would improve residents' quality of life. The city would thus become more attractive to new businesses, especially advanced technology firms and the professionals they employ. This revised conception of the rivers repositioned them as a key component, once again, of the region's economic development strategy. However, as this view became generally accepted, a new tension surfaced between environmentalists and municipal boosters. Appeals for the restoration and preservation of river ecosystems conflicted with development proposals that sought to exploit the region's recreational and tourism potential.

In the late 1970s, earlier predictions of absolute decline in Pittsburgh's industrial base began to come true. By the early 1980s, older steel mills, factories, and coal mines were permanently shuttered, leaving the region in crisis. Seemingly all at once, thousands of acres of formerly vital industrial riverfront turned into vacant and often polluted brownfields, commercial river traffic plummeted, and river infrastructure fell into disrepair. Under intense pressure to replace the many thousands of lost manufacturing jobs with new sources of employment, civic leaders cast about for all kinds of development opportunities and governmental support. Pittsburgh's weakened economy and Rust Belt image, the real and perceived toxic condition of brownfields, the outdated character of the industrial areas' highway networks, and the lack of a unified development strategy among the region's many local governments stymied new investment, especially along the older industrial riverfronts.[31] Ironically, this failure saved these critical sites from the prospect of redevelopment without adequate consideration of the rivers' potential in the "new" Pittsburgh.

Although generally greeted by civic leaders with polite indifference, several studies by governmental agencies and nonprofit organizations in the 1980s explored the question of how the rivers might be used to improve the region's quality of life. Most Pittsburghers dismissed the recommendations of these reports as impractical and elitist because of the region's cascading underemployment and unemployment that spelled hard times for thousands of families. Nevertheless, a few new waterfront projects indicated that the rivers might become popular with the public and eventually with investors, just as other such developments had already demonstrated in Baltimore, Portland (Oregon), and several other American cities. A small public boat-launching ramp and linear park along the city's South Side riverbank attracted hordes of weekend boaters. The new, high-profile annual River Regatta, featuring Formula One speedboat races, drew large crowds. A marina, riverfront park, and hike-and-bike trail similarly enlivened the depressed industrial city of McKeesport several miles south of Pittsburgh, while a waterslide amusement park flourished on the edge of a riverfront brownfield in run-down West Homestead, which had formerly housed a massive industrial-machinery works. Finally, a small sculpture park and office development called Allegheny Landing spruced up the riverfront across from downtown on that river's north shore.[32]

By the early 1990s, the attitudes of investors and public officials toward the development prospects of the riverfronts for scenic and recreational opportunities had clearly shifted. The successful redevelopment of an industrial island in the Allegheny River (Herr's Island, renamed Washington's Landing a couple of miles above downtown) into a complex of offices, light industry, a large marina, and townhouses with a circumferential walking and biking trail overlooking the water showed how important the rivers were to the region's future economic growth and stability. Recreational boating along with the attendant marinas and yacht clubs proliferated throughout the river valleys. The revival of rowing added another dimension to river participation, while river tours and commercial party boats attracted other clienteles. Most importantly, civic leaders and government officials embraced the idea of turning the north shore of the Allegheny River across from Pittsburgh's downtown area into a signature development dependent on the river ambience. The North Shore Corridor, as it has been called, became the site of two new baseball and football stadiums oriented to the river with views of downtown. The Aluminum Company of America (ALCOA) built its new headquarters along the river, where office buildings, entertainment venues, retail shops, and private residences have since been developed with support from the city. On the south (downtown) bank of the Allegheny, the newly rebuilt and enlarged "green" convention center was oriented to face the waterfront. The adjacent cultural district that has been emerging for more than twenty years contributed to the North Shore Corridor with new buildings, signage, and lighting as well as a linear park along the river's edge. A group of civic leaders and foundations formed the Riverlife Task Force, which conceived of a Three Rivers Park consisting of public spaces along the city's central riverfronts. The Riverlife Task Force has taken it upon itself to steward the development of these public spaces, coordinating with private developers and public agencies as well as negotiating for appropriate designs.[33]

Not all development of the brownfields has taken advantage of their location alongside the rivers. For example, the new development on the former site of Homestead Steel Works, which stretches for nearly three miles along the Monongahela about eight miles from downtown Pittsburgh, generally ignores the river despite being named "The Waterfront." The developers designed the complex as a suburban-style retail, restaurant, and office mall; only the walking trail on the riverbank and a few townhouses and other buildings face the river.[34]

The collapse of industry and the diminution of associated pollution and traffic from the rivers, along with state and federal clean-water regulations, have markedly improved both water quality and the river ecology. Biologists have determined that in the past "the upper Ohio River drainage basin was so grossly degraded that these systems were completely devoid of fish life."[35] The ecological recovery has been surprisingly rapid and extensive. Fish have returned to the rivers. In one recent study of tributaries of the three rivers, twenty-nine species of fish were discovered in streams flowing to the Allegheny, sixteen species in tributaries of the Monongahela, and thirty-two species in a tributary of the Ohio River. This significant recovery had sports enthusiasts, environmentalists, and Pennsylvania State Fish and Boat Commission employees cooperating to bring the Bass Master Classic fishing tournament to Pittsburgh for the first time.[36]

The woody vegetation along the region's riverbanks was also severely degraded, but as with the fish, woody plants have made a vigorous return to the rivers' once-bare and eroded edges. A recent study of the Monongahela and Allegheny rivers identified the existence of four native woody-plant communities and one native herbaceous-plant community typical of large rivers in North America. It also determined that the frequency of invasive species decreases with distance from the city's Point.[37] This rapid regeneration of riparian vegetation has taken observers by surprise. Organizations concerned with riverine ecology such as the Friends of the Riverfront, the Pennsylvania Environmental Council, 3 Rivers 2nd Nature, and the Western Pennsylvania Field Institute held several workshops to discuss the creation of water trails and in the process discovered several local sites of intense natural diversity. For instance, Sycamore Island, a large wooded island in the Allegheny River nine miles from downtown Pittsburgh, has few sycamore trees today, but its silver maples are so large and so tall that people seeing photographs of these maple groves assume that the setting is a national forest.

The ability of the rivers and streams to support life and the return of typical riparian vegetation reveal a natural potential that has been dormant for more than a century and suggest the possibility for human intervention to aid the process. Planning is under way for the restoration of natural riverine systems at several regional sites. The U.S. Army Corps of Engineers, for example, is engaged in the process of restoring an urban stream called Nine Mile Run. This stream flows from creeks that have long been directed into sewer pipes and from springs in a city park through a three-hundred-acre reclaimed brownfield site into the Monongahela River.[38] Additional plans and activities are under way to restore, conserve, and preserve local ecosystems. Recent river restoration studies by the U.S. Army Corps of Engineers, island studies by the U.S. Fish and Wildlife Service, and land conservation along the rivers by the Allegheny Land Trust reveal the gradual shift in thinking from traditional conservation of the hinterlands to conservation and restoration of land along the region's most industrialized waterways.[39]

Ecological restoration is difficult despite the remarkable recovery of natural systems. Riverine ecological systems encounter wet-weather threats to water quality from outdated sewage systems; the Pittsburgh region has the highest number of combined sewer outfalls in the country. Recurring episodes of mine-acid drainage also endanger ecological rejuvenation. Furthermore, some of Pittsburgh's forested hills and slopes near the rivers are once again being considered for rezoning to spur development. There has also been a proposal to construct an expressway through four Monongahela waterfront communities, which would affect two miles of reforested riverbank—the longest undeveloped stretch within the city limits. Clearly, the old development values still tug at Pittsburghers, even as natural systems have begun regenerating in the brief twenty-year respite from intense industrial use.

The ecological recovery has nevertheless spawned a growing interest in the aesthetics of river viewing. Public spaces with panoramic river views from walkways, bridges, hills, parks, and the water itself create outstanding opportunities to enjoy the rivers, their recovering banks, and the forested hillsides (figure 4.5). Real estate developers are erecting townhouses with water views as a means of capturing the rising economic value created by this aesthetic. This ecological recovery thus enhances the city's intention to become known as a place where residents and visitors are able to play in, on, and along the rivers and riverbanks.

The postindustrial view of nature emphasizing preservation and restoration of riverine ecologies, however, clashes at times with development that leverages the rivers' recreational and scenic attributes. Intensive housing and office complexes, manicured landscaping of banks, and noisy, high-speed boats roiling river waters all contrast with the aesthetic of ecological preservation. Rowers and power boaters uneasily share the water; birders despair at the managed vegetation of developers' complexes; and commercial shipping still has navigational priority.

The concepts of the environment as an amenity and as an ecological aesthetic both depend on a physical, sensual relationship to place, appealing to fashionable values of youth and an active, healthy lifestyle. Clean rivers with green edges and trails, it is argued, support this kind of lifestyle. Both concepts rely in distinct ways on continued management of the three rivers, and despite their inherent contradictions, the combination of these concepts has become critical environmental infrastructure that is essential for Pittsburgh to compete with other regions.

Urban Rivers

Throughout the city's history, Pittsburgh's succession of civic leaders has defined the rivers with respect to the three periods of its economic life. These periods conform generally with the experiences of other American river cities. Although the timing and specifics vary, Pittsburgh's river story also resonates with those of many European cities. The experiences of older industrial cities such as Newcastle-on-Tyne in England closely parallel those of Pittsburgh, but even Paris's relationship with the Seine, as suggested by Isabelle Backouche elsewhere in this volume, shares many of the same general characteristics.[40] Like Pittsburgh's three rivers, the Seine supported multiple functions and the "ordinary" life of Parisians until the late eighteenth century. After 1760 business elites and government officials increasingly regulated and reshaped this famous river according to a "mono-functional" vision of it as a national trade artery in an industrializing economy. Slowly the Seine became a stranger to residents, writes Backouche, though not to tourists. Finally, Parisian leaders in recent years have tried to reconnect the river to the lives of citizens in hopes of again making the Seine part of Paris's public space.

In some ways the segments of rivers that flow through urban areas shared many characteristics with their country cousins. Farmers, loggers, millers, and country merchants also regarded rivers as natural resources critical to their economic enterprises; if needed be, these rural entrepreneurs endeavored to regulate and manage them with, for example, dams or diversion canals. These rivers generally were not as intensively managed and exploited as the urban segments, where greater waste loads, hardening of edges, and filling or "reclamation" often drastically altered the ecology.[41]

Such intense intervention in urban rivers readily invites a historical interpretation that emphasizes the now-familiar dichotomy between nature and culture.[42] While now discredited in academic literature, this dichotomy was clearly reflected in public perceptions. Until recently the heavily engineered three rivers did not seem like "real" rivers to most Pittsburgh residents. Just as the Seine became a stranger to Parisians, the three rivers became inaccessible and unusable to ordinary Pittsburghers. Human activity had so degraded both water quality and riparian flora and fauna in pursuit of urban industrial development that most citizens had ceased to think of the rivers as part of nature. As infrastructure, the rivers might as well have been buried in

Figure 4.5. The peninsula at the confluence of Pittsburgh's three rivers, known as the Point.

culverts or concrete channels to do their job. In the search for nature's scenic and recreational amenities, Pittsburghers left the city for rural rivers or more distant sites of recreation. This industrial nature was "unnatural."[43]

Today the ecological rejuvenation of the three rivers and their rediscovery as part of the public commons depends on continuing regulation and management by those with the power to make policy, effect development, and enforce laws. The return of the rivers to their preindustrial state is, of course, no less a cultural construct than was the riverine ecology during the period of the rivers' industrialization. As environmental historian William Cronon has noted, nature "is a profoundly human construction . . . the way we describe and understand [it] is so entangled with our own values and assumptions that the two can never be fully separated."[44] The restoration of a preindustrial ecology will be only partially achievable. The maintenance of industrial-era infrastructure such as locks, dams, and hard edges; the alterations caused by dredging for sand and gravel; and the presence of invasive species will prevent complete restoration. Furthermore, the pressure for development, even of some river-friendly recreational and residential projects, will introduce highly managed landscapes. Nonetheless, with riverfront trails; increased boating, rowing, and fishing; and the renewed pleasure of river viewing, the three rivers, as reconceptualized and shaped by humans, are again part of the public commons and critical to Pittsburgh's reinvention of itself.

5

The Cultural and Hydrological Development of the Mississippi and Volga Rivers

DOROTHY ZEISLER-VRALSTED

The story of the Moscow-Volga Canal and the construction of the Upper Mississippi River locks and dams offers a comparative case study of "highmodernist" thinking on the environment.[1] Both projects reflect the prevailing attitude toward natural resources in the 1930s—namely, that such resources could and should be shaped to better serve an industrialized society. In addition, both projects were considered examples of cutting-edge engineering. Among the many perceived benefits of the Moscow-Volga Canal were the provision of drinking water for Moscow's rapidly growing population, easy access to the Volga from the city of Moscow, and the modernization of preindustrial villages along the canal route. In the United States, the Army Corps of Engineers used "technological breakthroughs in navigation engineering" to create an intracontinental channel that permanently transformed the Upper Mississippi into a commercial highway.[2] The Mississippi River project also helped midwestern farmers to remain competitive in the domestic and international agricultural markets.

The ecological consequences of these projects—which include degraded water quality, elimination of the natural floodplain, bank erosion, the disappearance of side channels and backwaters, as well as the loss of species found in those habitats—are still being realized. But at the time these developments were being planned and executed, they were celebrated as symbols of both modernization and a successful industrial nation-state.

For many Western nations the decade of the 1930s was an era of massive public works projects fueled by the conviction that science and technology could transform and ultimately control the environment. As a result, studies of Franklin Roosevelt's New Deal and Joseph Stalin's Five-Year Plan have focused on how this era's blind faith in industrialization has affected human populations. There are, however, other actors to consider, such as the rivers and streams that were radically altered by the creed of progress and the public works it promoted.

During this period the Volga and Mississippi rivers were drastically modified, but despite very different political ideologies and administrative processes, the outcome in both cases was similar. By 1940 the upper reaches of these two rivers were no longer free-flowing waterways; each had become a series of slack-water ponds. Thus, from an environmental perspective, the divergent political rhetoric supporting modernization by means of public works proved to be insignificant, since the impact on the natural resources in question was essentially the same. In other words, the drive to modernize superseded political ideologies.

In the 1930s, however, American and Soviet promoters of public works championed the political systems that realized these projects. When architects of New Deal planning, such as David Lilienthal, praised the Tennessee Valley Authority (TVA), he emphasized that this massive project was the product of a democratic society. Russian leaders, too, when boasting of their nation's engineering triumphs, associated them with the supremacy of Soviet ideology. In Maxim Gorky's journal *Our Achievements,* the accomplishments of the first Five-Year Plan were publicized with claims that in the Soviet Union, "men were no longer the helpless victims of economic forces over which they had no control."[3] These products of the "high-modernist" ideology that prevailed during the period reflected growing "self-confidence about scientific and technical progress . . . and the rational design of social order."[4] High modernism transcends local interests in natural resources conservation and enables governments to undertake large-scale development.

Even before the modern period, however, both rivers were central to a national narrative. Each river served as a major transportation artery, and each was idealized in literature and the visual arts. The folklore surrounding these waterways is rich with references to "Mother Volga" and "Old Man River." Contributing to their iconic status are their size and topography. Stretching 3,705 kilometers from its source in northern Minnesota to its mouth in the Louisiana Delta, the Mississippi is the longest river on the North American continent. Even more remarkable than its length is the size of its drainage basin, which includes 4.76 million square kilometers, making it the third largest watershed in the world. The river begins in the north woods of Minnesota, winds through midwestern farmland, and empties into the subtropical Gulf of Mexico. The habitats the river supports are equally diverse, and all of them sustain numerous species of fish and wildlife.[5]

The Mississippi is divided into three major sections: the headwaters, the Upper Mississippi River basin, and the Lower Mississippi River. The headwaters begin at Lake Itasca and flow southward past Minneapolis. There the Upper Mississippi River basin begins and runs downstream to Cairo, Illinois, where the Ohio River empties into the Mississippi. This junction marks the beginning of the Lower Mississippi, which ends in the Gulf of Mexico. In addition to the Ohio, the river has one other major tributary, the Missouri, which joins the Mississippi near St. Louis.

The Volga is the longest river in Europe, with a length of 3,700 kilometers. Starting in the Valday Hills (situated northwest of Moscow), it flows in a southeasterly direction and empties into the Caspian Sea near Astrakhan. With 151,000 rivers, brooks, and streams emptying into the Volga, its watershed covers an area of 1.45 billion square kilometers or 8 percent of the country's territory. Of the 2,600 rivers that flow into the Volga, the principal tributaries include the Kama, Samara, Oka, and Vetluga rivers. The terrain of the river basin ranges from a forested zone of southern taiga in the north to forest-steppe to the arid delta area in the south. The river basin supports habitats that sustain seventy-four species of fish and a large commercial fishing industry.

Like the Mississippi, the Volga is divided into three parts—upper, middle, and lower. The Upper Volga's source is near the village of Volgaverkhovye in the Valday Hills, and this portion of the river extends eastward to the Rybinsk Dam, where the Sheksna River joins the Volga. Below the Rybinsk Dam the Middle Volga flows to the Kuibyshev Reservoir, where the Lower Volga begins and runs southward until it reaches the Caspian Sea.

While no major changes were introduced to these rivers until the twentieth century, both were viewed in the eighteenth and nineteenth centuries as critical transportation routes for growing economies. Paralleling their commercial value was a visual and print culture that sought to capture the aesthetics of each river. Beginning in the 1840s, several artists painted panoramas of the Mississippi, with the first well-known work being *Banvard's Panorama of the Mississippi, Painted on Three Miles of Canvas, exhibiting a View of Country 1200 Miles in Length, extending from the Mouth of the Missouri River to the City of New Orleans, being by far the Largest Picture ever executed by Man* (c. 1840). The next important example of this genre was the four-mile-long panorama painted by Henry Lewis, which he referred to as the "Great National Work." Unlike Banvard, Lewis included part of the Upper Mississippi in an area near Minneapolis. Both artists exhibited their panoramas to audiences in the eastern United States and in Europe. Three other major panoramas were also painted during this period.[6]

The steamboat inspired more visual representations, with images of such vessels docking or floating along the Mississippi soon becoming familiar fare. Tours of the river also became popular after public figures, such as Henry David Thoreau, began traveling the river by steamboat. In 1854 the Upper Mississippi was celebrated with a public relations campaign developed by two railroad financiers, Henry Farnam and Joseph Sheffield, who had built the Chicago and Rock Island Railroad. Their "Great Excursion," as it was known, began at Rock Island, Illinois, and included twelve hundred guests, many of them renowned figures, such as former president Millard Fillmore and Charles A. Dana of the *New York Tribune*. Praise for the scenery came from several of the passengers, and one compared the Upper Mississippi to the Rhine River. But the most glowing description came from a "special correspondent" to the *New York Daily Times*, who wrote, "Perhaps you have beheld such sublimity in dreams, but surely never in daylight walking elsewhere in this wonderful world. Over one hundred and fifty miles of unimaginable fairy-land, genii-land, and world of visions, have we passed during the last twenty-four hours. . . . Throw away your guide books; heed not the statements of travelers; deal not with seekers after and retailers of the picturesque; believe no man, but see for yourself the Mississippi River above Dubuque."[7] The Great Excursion inaugurated "boom times in the Upper Mississippi Valley" before the railroads dominated transportation in the Midwest.[8]

Accompanying the tours and pastoral images was a growing folklore of the exploits of various river pilots or the earlier keelboat men, such as the legendary Mike Fink. Beginning in the 1830s, river folklore echoed the same themes that Mark Twain would later underscore in *Life on the Mississippi*, in which he portrayed the river men as mythological figures famous for their independence, their love of the great waterway they traveled, their tendency toward brawling, and especially their navigational skill.[9] Experienced pilots were needed because of the river's unpredictability and tendency to create new channels, "straightening and shortening itself . . . "thirty miles at a single jump," Twain wrote.[10] In response to the river's changeability, the Army Corps of Engineers began "improvements" on the river in the 1830s. Initially, Corps engineers "worked to remove snags, sandbars, and shoals" in order to accommodate increasing commercial and passenger traffic. Later projects included a four-and-a-half-foot channel and a series of wing dams, with more than a thousand constructed between the Twin Cities and La Crosse, Wisconsin (a distance of 140 miles) by 1930. Despite attempts to facilitate navigation, however, the Upper Mississippi remained a challenge for river pilots; not until the 1930s, when the locks and dams were built, did the river become the commercial highway it is today.[11]

The Volga also has a long tradition in Russian folklore. One recent interpretation of Russian attitudes toward nature and their landscapes, by Christopher Ely, contends that Russian intellectuals had to learn to appreciate their landscape, in part, because their aesthetic sense, like those of American artists and audiences, had long been dominated by European conventions of landscape painting. When Russian intellectuals began to reexamine their native landscapes in the nineteenth century, the Volga was a predominant force. For example, in 1838 the Russian government commissioned two artists to float down the Volga and "document the beauty of the great river." Poets also championed the river; as Christopher Ely notes, "Aleksandr Sumarokov, Ivan Dmitriev, [and] Nikolai Karamzin all wrote poetry in praise of the Volga as a symbol of Russia."[12]

One of the best-known Russian poets to write about the river was Nikolai Nekrasov, whose work appeared in the second half of the nineteenth century. Nekrasov, who grew up along the Volga, captures its beauty and its place in Russian culture in such poems as "On the Volga River," in which he praises its unchanging majesty: "I've changed a lot, but you are the same. / So light, so majestic, as you used to be."[13]

Landscape artists such as Isaak Levitan also celebrated the river. Levitan spent many summers in the village of Plyoss on the banks of the Volga, where he created some of his best landscapes, such as *Evening on the Volga* (1888). Levitan also exhibited another aspect of Russian exceptionalism in his representation of space and the immense, unbroken Russian plain, which was often a backdrop to the river in his paintings. Rivaling Levitan's works are the paintings by Aleksei Savrasov, especially his *End of Summer on the Volga* (1873). The riverscapes presented by both of these artists reflect a serenity and tranquility seen in many representations of the Mississippi River.

Among the best-known paintings of the Volga is Ilya Efimovich Repin's *The Volga Boatmen* (1872), which refers to another theme in the river's narrative. Before the advent of steam-powered vessels, transport on some parts of the river was provided by bargemen, or *burlaki*, many of whom were runaway serfs. Echoing the folklore of the keelboat men on the Mississippi, the *burlaki* were often characterized as free spirits who were more comfortable on the social periphery. Nineteenth-century painters captured this outsider ethos in their depictions of the Volga and Mississippi rivers and these rivers' major tributaries. George Caleb Bingham pictures an easygoing life on the river in such works as *Jolly Flatboatmen* (1846) and *Raftsmen Playing Cards* (1847), and in his paintings the marginal status of watermen is portrayed as enviable rather than pitiable. In contrast, Repin shows the viewer that a bargeman's existence was a harsh one.[14] The poet Nekrasov echoed this bleakness of the bargemen's lives: "Along the river there were barge-haulers, / And their funereal cry was unbearably wild."[15] Whatever differences there may have been between these portraits of East and West, each river contributed to the emerging national narratives of the nineteenth century that celebrated the exceptionalism of their respective landscapes.

The first Russian leader to propose an engineering project that would improve commercial shipping on the Volga was Tsar Peter the Great, who conceived the idea of building the canal between Moscow and the Volga. Such a waterway would make it possible to transport goods by water from Moscow all the way to the mouth of the river. After Peter's death, although smaller projects were undertaken to improve navigation, the tsar's initial plan was not raised again until the late 1920s under Joseph Stalin. The need for a canal from Moscow to the Volga temporarily disappeared when freight train routes were established, but the canal was nonetheless built during Stalin's rule to supply water for domestic and industrial uses and to provide another avenue for shipping goods and resources. Upon completion of the canal, Moscow became a port city with access to four seas—the Caspian, Black, Azov, and Baltic. In 1931, when construction of the canal began, sufficient rail capacity did not exist to meet the needs of what was becoming a major industrialized nation. Thus the canal was built to supplement the railroads and provide a cheaper form of transportation.[16]

By contrast, the construction of a nine-foot intracontinental channel on the Upper Mississippi was initiated because of the railroads' monopoly on freight traffic. In the years preceding the Civil War, the river served as an important commercial highway, carrying lead ore from mines in Illinois and Wisconsin and later transporting immigrants to the Upper Midwest. Even after the arrival of the railroads, rail and river traffic were interdependent for several decades, as newly built railway lines intersected with steamship routes. Following the Civil War, production and shipping increased dramatically, and many questioned the growing reliance on railroads. By the 1870s, however, the need for greater carrying capacity was not the primary issue sparking concerns about the railroads' monopoly on transport; instead, farmers and local businesses were dissatisfied with the high fees the railroads could command. But regardless of the motives for improving navigation on both the Volga and the Mississippi rivers, the outcomes were similar—each river became the nation's highway. The reservoirs, locks, and dams needed to create these transportation arteries did, however, incur significant environmental and social costs.[17]

From the mid-1860s, farmers, flour-mill operators, and related business interests in the Upper Mississippi River basin sought ways to improve navigation in order to make shipping by water more cost effective. By the end of the nineteenth century, river trade had dwindled to the point that the only major shipping consisted of the log rafts of timber cut from the northern forests of Minnesota and Wisconsin. Prior to this, the principal commodity shipped on the river had been grain from Minneapolis, a city renowned for its booming flour mills. In an effort to accommodate the needs of the Minneapolis grain market and to provide another competitive means of transportation, the Corps of Engineers in 1878 began to build a four-and-a-half-foot channel. It soon became apparent that a deeper channel was necessary, and the Upper Mississippi River Improvement Association was formed. In contrast to development of the Volga, the engineering of the Mississippi was prompted by local demand. Each significant project on the river was initiated by local organizations, often consisting of business leaders who shipped grain, operated barge lines, or owned mills. Their concerns about the decline in river traffic were well founded, as commerce on the Upper Mississippi River fell 85 percent between 1889 and 1906.[18]

Lobbying by the Upper Mississippi River Improvement Association led to congressional authorization of a six-foot channel in 1907. The early years of the twentieth century looked promising for the Mississippi, with political leaders such as President Theodore Roosevelt remarking on the importance of the Mississippi River Valley and its central role in America's future. Funding for the six-foot channel was limited, however, and by the mid-1920s the project was still not complete. By this time, promoters and engineers had realized that a six-foot channel would no longer be sufficient and had focused their energies on a nine-foot channel instead.[19] The federal government also supported a nine-foot channel project on the Upper Mississippi River, as did farmers who depended on the sale of grain for their livelihoods and who had to this point been wholly reliant on the railroads for shipping. By opening up another viable transportation route, farmers and business leaders reasoned, shipping rates would become more competitive. Supporters also stressed the potential for an expanded international market.

The nine-foot-channel project was not without its detractors, who included the conservationists of the Izaak Walton League and Major Charles Hall, one of the Corps engineers charged with evaluating the proposal. Smaller river towns showed little interest in the project, and reasons for opposing it varied. The Izaak Walton League had succeeded in establishing a wildlife refuge in the Upper Mississippi River Valley in 1924, and members of the league viewed the proposed channel and the increased traffic and pollution it would unquestionably bring about as a threat to the refuge. Major Hall's chief criticism was that the cost of the project would exceed the anticipated commercial gains. While the channel did have detrimental effects on the environment, Hall's criticism proved unfounded. Barge traffic was so much greater than expected that business interests in the Upper Mississippi River Valley were soon calling for a twelve-foot channel.

Despite the opposition, which in itself was unusual since earlier navigation projects had encountered no criticism, construction of the nine-foot channel was authorized in the River and Harbor Act of July 3, 1930. President Hoover signed the act but did not award full funding. Promoters of the project persisted, however, and in 1933, 1934, and 1935 additional funding was provided by the Public Works Administration. Deepening of the channel thus became a public-works project with a mandate to provide jobs during the worst years of the Depression.[20]

As with other New Deal developments, the Corps documented every phase of the channel construction. Photographs were taken, along with film footage for technical and nontechnical viewing. Altogether, Corps personnel shot almost "19,000 feet of film with almost 14,000 feet related to actual dam and lock construction." A monthly journal titled *Old Man River* was published with project updates and safety tips for the workers. Unlike supervisors of the Moscow-Volga Canal construction, who exploited gulag labor, Corps personnel worried about the safety of so many unskilled workers. But despite their precautions, lives were lost: in a single district there were eleven deaths due to accidents. The locks and dams were products of engineering technology that in the 1930s (considered the Corps's "Golden Age") was still new. Originally developed to build the "non-navigable dams that incorporated both roller and Tainter gates," this technology was obsolete by the time the final lock and dam were completed in 1940.[21]

Complementing and preceding the public works projects of the 1930s, however, was a broader movement in politics and culture to manipulate and transform water resources. Major construction such as the Panama Canal at the turn of the twentieth century and numerous reclamation projects had been designed to serve multiple purposes, and they contributed to a culture shaped by the language of development. One of the first projects to embody the multipurpose concept of dam building was the Boulder Canyon project, which resulted in the construction of Hoover Dam. With the authorization of this federal project, multiple-purpose planning became the standard. New projects combined navigation, energy production, flood control, irrigation, and other potential benefits. The best example of the shift to large-scale projects is the TVA, a New Deal project intended to provide not only the aforementioned benefits but also an electrical power supply for residents of rural areas. Although the American process of developing water resources often differed from that of the Soviet Union, especially with its reliance on gulag labor and Stalin's occasional disregard for the views of the engineering community, projects such as the TVA and the proposed Columbia Valley Authority had goals that were similar to those promoted in Soviet projects.[22]

Throughout the 1930s the engineering community in the United States—whether through the bureaucracy of the Public Works Administration, U.S. Reclamation Service, or Army Corps of Engineers—dominated water projects. This was the age of the expert, and well-known engineers, such as Elwood Mead, regarded rivers as resources waiting to be developed. Leaders in the Soviet Union expressed comparable views regarding nature. Leon Trotsky argued that "through the machine, man in socialist society will command nature in its entirety. . . . He will point out places for mountains and passes. He will change the course of the rivers and will lay down rules for oceans."[23]

Locks and dams, hydropower stations, and irrigation work were thus all symbols of modern scientific expertise. In the 1930s these symbols often had broader meanings that suggested a brighter economic future, to which President Franklin Roosevelt alluded when he visited a project on the Upper Mississippi and spoke of the "possibilities of this upper valley."[24] The best example of this faith in the potential of water projects can be found in David Lilienthal's comments on the power of the TVA as an agent of social change: "A great Plan, a moral and indeed a religious purpose, deep and fundamental, is democracy's answer both to our own homegrown would-be dictators and foreign anti-democracy alike. In the unified development of resources there is such a Great Plan: the Unity of Nature and Mankind. Under such a Plan in our valley we move forward. . . . But we assume responsibility . . . [for] the material well-being of all men and the opportunity for them to build for themselves spiritual strength."[25] Roosevelt, too, saw in these large-scale projects the promise of a better life for America's rural poor.[26]

Thus in the 1930s the language of development and its visual representations had far-reaching social ramifications that persist even today. To the champions of industrialization in the United States and the Soviet Union, modernization signified more than technological advances. As early as the turn of the twentieth century, scientists and boosters alike recognized that technology—whether for irrigation, flood control, hydropower, or navigation—could improve the human condition. In the Soviet Union, the Bolsheviks viewed rapid industrialization as a goal long before they sought revolution; by the end of the first Five-Year Plan under the Communist regime, the Dnieper River hydroelectric dam had been completed. To Stalin industrialization meant a break with a Russian past that had resulted in defeat at the hands of developed nations. In his words, "All beat her [Russia]— because of her backwardness.... We are fifty or a hundred years behind the advanced countries. We must make good this distance in ten years. Either we do it or we shall go under."[27]

Americans also worried about falling behind. In arguments for the nine-foot channel on the Upper Mississippi, advocates stressed the importance of staying competitive in the international grain market. From a broader perspective, proponents of New Deal projects touted the benefits of modernization (especially in the form of public power) for the rural poor in places such as Appalachia. Massive undertakings like those on the Volga and Mississippi rivers constituted a break with a rural, backward past, whether by transforming rural villages along the Volga or by creating an expanded market in the Upper Mississippi River Valley. This is not to say that differences between the two projects did not exist—even the engineering of both was in many ways dissimilar as a result of differences between command and market economies and contrasting political ideologies. Yet in many instances the motivations were the same, with similar rhetoric and environmental consequences.

The construction of the Moscow-Volga Canal, extending from Moscow to Dubna for 128 kilometers and including eleven locks and dams, was part of Stalin's second Five-Year Plan. Today the history of the canal has become a large part of Russia's national narrative due in part to the accessibility of new archival sources and recent scholarship that has exposed its "darker" legacy. Contemporary accounts reveal that planners initially considered several routes, and when a plenary session of the Central Committee of the VKP (All-Union Communist Party of Bolsheviks) resolved on June 15, 1931, to build the canal, the route had still not been determined. Although the stated goals of the canal were to improve navigation and to provide Moscow with water for domestic and industrial uses, Stalin reportedly had another goal. According to one source, the canal was an opportunity for him to promote himself and to show there "was nothing that communism could not do." Canals interested Stalin and, in the words of one scholar, "seized his imagination . . . it sometimes seemed as if he wanted to dig them almost indiscriminately." Before the Moscow-Volga project, Stalin's only major water project had been the White Sea Canal, which connected the White Sea with Baltic ports.[28]

In Stalin's ambition to transform the Soviet Union into an industrialized nation, he relied on the labor of thousands of prisoners to build the White Sea and Moscow-Volga canals. Many of those who worked on the construction of the Moscow-Volga Canal, which officially began in 1932, had been imprisoned for minor "political crimes," such as the telling of an anecdote. In a letter written in 1934 by Genrikh Yagoda, who headed the project, he requested fifteen thousand to twenty thousand prisoners, saying they "were needed urgently in order to finish the Moscow-Volga Canal." These prisoners worked at breakneck speed, often with wheelbarrows and shovels, to finish the canal and completed it in four years and eight months. As a result, in 1937, the first ships traveled up the canal to Moscow.[29]

While the use of forced labor distinguished the Moscow-Volga Canal's construction from that of the Upper Mississippi River locks and dams, both projects generated a rich visual culture that publicized their presence and meaning. Artists were drafted into service to promote the Moscow-Volga Canal, and the many images they produced include representations of the merits of industrialization, portrayals of the nobility and dignity of the worker, and drawings of well-known political figures. The person responsible for most of this visual record is Syemyen Firin, the chief administrator of the prison camp in Dimitrov. Credit belongs to Firin for discovering the best-known artist of the canal, Gleb Kun. Firin stumbled upon Kun (initially a laborer) drawing one day instead of working on the canal. Rather than punishing him, Firin appointed him chief artist of the project. In his new role Kun produced many works—in traditional Soviet style—celebrating this undertaking. He made sketches of women and men laborers as well as drawings of the dams and actual construction. In 1937, when the first ships traveled up the canal, all along the banks were Kun's portraits of Soviet leaders. Despite his talent and contributions, Kun and many of his colleagues were executed later that same year. The official reason for their executions was their role in a proposed plot to assassinate Nikolai Yezhov, then head of the Narodny Komissariat Vnutrennikh Del (People's Commissariat of Internal Affairs), or NKVD, the Soviet secret police.

Another consequence of the Moscow-Volga Canal and its accompanying dams was the relocation of 110 communities whose towns were inundated as part of the project. One of the better known of these towns was Korcheva, which was built by Catherine II and is still remembered today. The most lasting impressions of the canal are the accounts provided by the workers and prisoners who built it. One of these accounts tells of prisoners submerged in icy waters and drowned when cofferdams around them failed and were swept away. Everyday life in the work camps was brutal, and the workers saw reduced rations if the day's work quota was not met. Workers whose output dramatically exceeded their quotas, such as the bricklayer who in a single shift laid 40,578 bricks—the equivalent of eight railroad cars' worth—became celebrities.[30]

Almost as soon as the canal was completed, barge traffic on it was substantial, reflecting a level of commercial activity comparable to that on the Upper Mississippi River. Despite the human costs of its construction, the canal did become a vital trade route from Moscow to the mouth of the Volga, turning many Russian villages along the river into bustling cities. During the construction years, Soviet art promoted the canal by projecting the image of sleepy rural hamlets being transformed by industrialization. For example, in a work by Gleb Kun, painted in the early 1930s and entitled *Two Dimitrovs*, there are two images of the town. One shows the old church with its surrounding *kremlin* or fortress-like wall, while the other depicts a young man driving a truck and a steam shovel in the background. Since the town of Dimitrov is located on the canal route, this representation of progress and modernity visualizes the bright future that the completed canal promises for Dimitrov. (The belief that Stalin's Five-Year Plan transformed rural Russia was widely shared in the 1930s and 1940s and still has adherents today.) Second, all of the work on the canal was done with equipment manufactured in the Soviet Union—a source of pride for many. For example, the pumps used at the construction sites were built at fifty plants in the biggest cities throughout the nation, while engineers designed a new type of riverboat and special 150-ton cranes, which were also used in the construction. For the Soviet Union, the canal was just the beginning of the Volga's transformation. In 2004 there were eleven hydropower stations on the Volga and its major tributary, the Kama River. In the Volga River basin 716 reservoirs have been built, and they supply 13 percent of the basin's power facilities. Two of the largest projects are the Rybinsk and Kuibyshev dams, built in the 1950s. Because of these power sources, 45 percent of Russia's industry and 50 percent of its agriculture are located in the Volga basin.[31]

The Moscow-Volga Canal and the nine-foot-channel project on the Mississippi accomplished their primary goals of enhancing navigation and, in the Russian case, delivering a municipal water supply. But the environmental degradation of ecosystems in both rivers has been significant. The future ecological integrity of these waterways depends on responsible management through which commercial and recreational needs are carefully balanced against each river's ability to fulfill increasingly numerous demands.

Environmental concerns on the Mississippi include the continuing loss of wetlands. The populations of many native organisms are also declining. Increased shipping on the river probably accounts for the arrival of nonindigenous species, which now threaten some native species. In recent years the population of zebra mussels, a bivalve mollusk first detected in North American waters around 1988, has flourished and could seriously disrupt the Mississippi's aquatic ecosystems. Other causes of the river's loss of flora and fauna include industrial pollution, which has been ongoing. In the spring of 2004, the conservation organization American Rivers placed the Mississippi on its list of the nation's most endangered rivers.[32]

Still, robust traffic and bottlenecks during the high shipping season have prompted proposals to double the capacity of twelve locks on the Mississippi and Illinois rivers. If constructed, these locks could handle twelve-hundred-foot barge tows, eliminating the need to separate them for passage through the six-hundred-foot channels. Once again, advocates argue that the increased capacity will enable U.S. farmers to remain competitive in the international grain market. Debates between environmentalists and developers have been spirited, since environmentalists view the barge traffic as a "tax-subsidized transformation of the Mississippi into an industrial waterway."[33]

The Volga's ecosystem is experiencing similar threats, and scientists are becoming alarmed about the poor water quality caused by unregulated discharge of industrial wastes into the river. (In 2003, of the estimated 42 million tons of toxic waste produced in the Volga River basin, only 13 tons were managed and treated.) Another major source of pollution comes from agricultural by-products. Native organisms are under siege as recent findings of fish without fins attest. The hydropower stations that block access to upstream spawning grounds also threaten the fish population.[34]

In the historical development of the Mississippi and Volga Rivers, the decade of the 1930s was the turning point for both. During this period the forces of modernization overshadowed any other considerations with the demand for two "barge superhighways." The ideology of high modernism, with its cadre of scientists and experts all possessed by a moral certainty about the power and desirability of a manipulated environment, dominated political culture in both countries.[35]

Yet the transformation of these rivers could not have occurred without earlier perceptions of them as highways. Their commodification is not recent but has been a driving force behind long-term efforts to improve their navigability. As one scholar of the Mississippi River has observed, "the Upper Mississippi is defined more by the visions that made it a navigation channel than by its own natural character." There were, however, significant differences between the two major waterway projects. Impetus for the locks-and-dams project on the Mississippi came from the local populations, while it was "top-down" decision making that brought about the Moscow-Volga Canal. Motivated by a market economy, midwestern farmers and business leaders petitioned the federal government to fund channel improvements. Adding to the federal government's role in developing water resources was the reclamation activity in the American West that further demonstrated the need for government oversight of large-scale projects. Public works carried out during the Great Depression also ensured a role for federal agencies. As a result, the federal government, through its corps of experts, became more engaged in the nation's water politics. So, by the 1930s, despite very different political ideologies, the Soviet Union and the United States each undertook development of their major river, through a team of experts, supported by the state government.[36]

In the 1920s both nations shared similar views on modernization, industrial growth, and the potential utopias they could produce. Soviet political leaders and scientists saw their natural resources and emerging technologies as opportunities for "limitless development." American political leaders mirrored these views. Franklin Roosevelt, for example, saw the broad, sweeping changes that resulted from the TVA and supported the creation of similar projects such as the Columbia Valley Authority. In the drive to modernize and subsequently exploit water resources, electricity was a predominant symbol of modern, industrialized society. In the United States the electrification of private homes and an expanded domestic market of electrical appliances became expressions of modernity. Soviet authorities, beginning with Lenin, echoed these sentiments, which were also attested to by a rich visual culture, with murals associating the merits of electrification with communism or images of transformed Russian villages.[37]

But American and Soviet engineers also "shared faith in the ability of technology to change the face of nature for the better," as well as similar attitudes toward nature and, in particular, water resources. To the American and Soviet technocracy of the 1930s, rivers were a resource to be exploited, whether as sources of hydroelectric power or as commercial highways. The culture of modernity, with its own symbols and ethos, superseded political ideology in the 1930s. It also took a severe toll on the ecosystems of the Mississippi and Volga rivers.[38]

This comparative case study of the Moscow-Volga Canal and the Upper Mississippi River lock-and-dam project offers several conclusions. First, these models of development, like their larger counterparts such as the TVA, became exportable. In succeeding decades developing nations, including India, China, and Afghanistan, undertook major water development projects, particularly large multipurpose dams. Dam building became the currency of modernization, and, in the words of one political leader, societies seeking to reshape the environment were infected with the "disease of giganticism."[39] Second, the political leaders and engineers in both countries had similar attitudes toward nature, attitudes that transcended political ideologies. Third, and perhaps most important, transforming each river into a superhighway produced similar destructive environmental consequences—a caveat for all future water-development schemes.

6

River Diking and Reclamation in the Alpine Piedmont
The Case of the Isère

JACKY GIREL

During the period known as the Little Ice Age, waterways in the Alpine piedmont changed from single-channel rivers meandering over a large floodplain into braided, or anastomosed, channel systems. This morphodynamic transformation was brought about not only by the lower temperatures and increased rainfall that characterized this era of meteorological crisis but also by anthropogenic climate changes in the northern French Alps resulting from the deforestation of regional watersheds over three millennia.[1] The multichannel systems that evolved after 1550 were quite unstable; whenever the wetlands flooded, new channels developed, some of the islands within them disappeared, and others were created. Consequently, Alpine valley landscapes usually contained the following elements: functional and abandoned channels colonized by aquatic and semi-aquatic vegetation; islands formed by alluvial deposits and colonized, according to their age, by herbaceous, shrubby, or woody plant communities; marshes, mainly present below the alluvial fan, at junctions with tributaries; and wet meadows and plowed fields on terraces or ancient islands.

About the middle of the eighteenth century these valleys began to be viewed as ideal transportation routes and agricultural sites. Converting them from wetlands to habitable communities and productive farms would require controlling the flow of water and sediments. Water unsuitable for navigation or potentially hazardous to health or to crops had to be made usable, and sediments had to be effectively managed. To accomplish these goals, Italian and

French engineers undertook a series of diking and drainage projects that by 1850 had drastically altered many Alpine riparian landscapes.

The model that eventually became synonymous with a so-called "natural" Alpine hydrosystem was established in the Isère River valley, located in the Rhône-Alpes region of southeastern France (figure 6.1). The Isère exemplified the complex interrelationship between socioeconomic and environmental factors in the creation of these new landscapes. In this case study I show how this river was transformed from a volatile braided-channel system into a diked single-thread waterway flowing through the center of rich farmlands.

The Riparian Landscape of the Isère to 1800

The history of Alpine countries was marked by great instability during the French Revolution and the Napoleonic Wars, a political context that strongly influenced regional development. Part of the kingdom of Sardinia from 1720, Savoy was occupied by Spain from 1742 to 1749, when it reverted to Sardinian control. From 1792 to 1815 it was occupied first by French and then briefly by Austrian troops, after which it was again restored to the kingdom of Sardinia

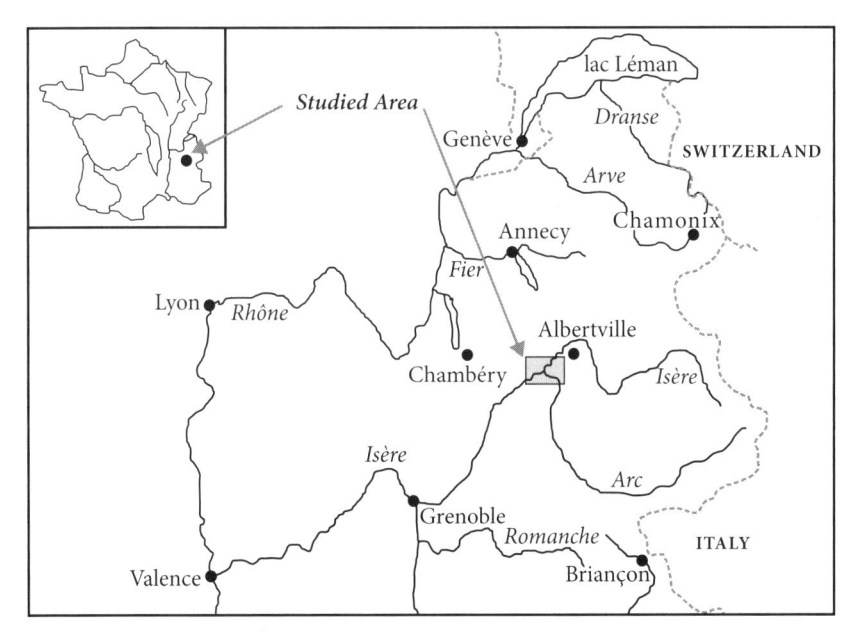

Figure 6.1. Hydrosystem of the Upper Rhône and Isère River valleys.

until 1860, when it was annexed by France. The riparian landscapes in Savoy thus reflected both the region's own legislation and policies as well as those of its neighbors. There are several contemporary depictions of the valley in question, the Combe de Savoie. One appears in the military map created in 1711 for Jacques Fitz-James, duke of Berwick.[2] Another is in Marchetti's map (1781) of the route from Chambéry to Turin, and a third is in the *Carte du Diguement de l'Isère et de l'Arc*, produced by Francesco-Luigi Garella between 1786 and 1789 as part of his survey (figure 6.2). These three documents show a low-sinuosity braided floodplain with numerous channels, islands, and such major structural components as alluvial fans, fens, and farmed terraces. Eighteenth- and early nineteenth-century land registrations, such as the Sardinian cadastre surveyed between 1729 and 1732 (the so-called "Mappe Sarde") and the Napoleonic cadastre surveyed between 1807 and 1810, along

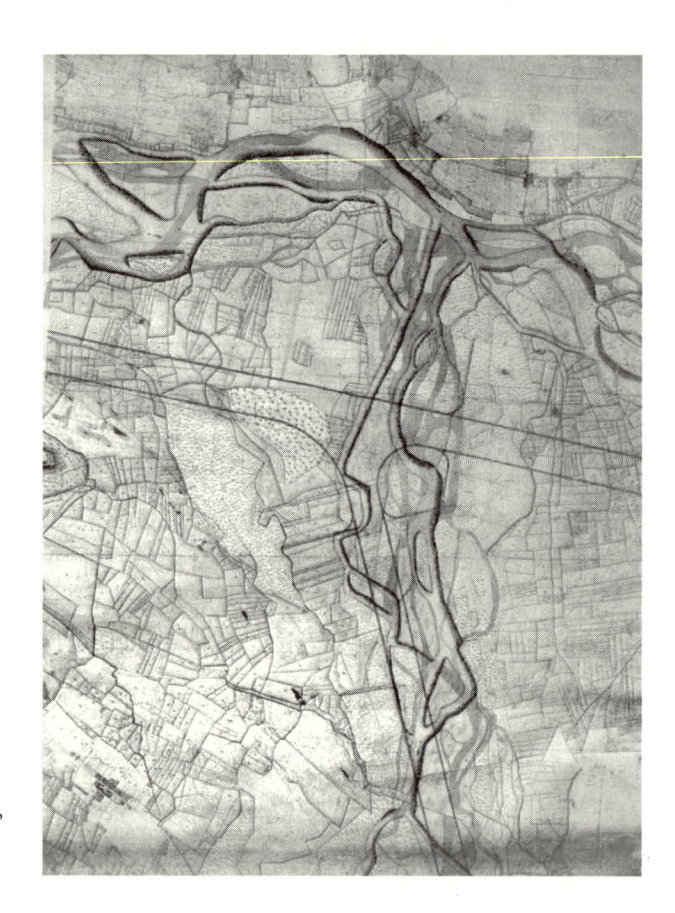

Figure 6.2. Francesco-Luigi Garella's map of the Arc-Isère junction (1786–1789), fol. 6 of the *Carte du Diguement de l'Isère et de l'Arc* (Archives Départementales de la Savoie, Chambéry, séries C&P, no. 1/13).

with contemporary manuscript records concerning land cover and land use, supply precise knowledge of the Combe de Savoie before it was developed by the French engineers of the École Nationale des Ponts et Chaussées.[3]

The presence of large fenlands—used as grazing commons, mowed meadows, and plowed fields—shows that the floodplain was heavily exploited despite frequent flooding and erosion. Only permanently flooded marshes were considered unproductive and therefore classified as insalubrious following the passage of a law on September 16, 1807, requiring that they be drained for cultivation.[4] The major part of the floodplain belonged to the parishes and was held in common. From it came the reeds, rushes, sedges, and grasses that were used as fodder for livestock and thus converted into manure, which was also an important component of the agricultural ecosystem. These various plants, known collectively as *blache,* were harvested annually at the end of August; the land was then used for cattle grazing and mowed at the beginning of the cold season. Small spiny trees, such as hawthorn (*Crataegus monogyna*), blackthorn (*Prunus spinosa*), and common barberry (*Berberis vulgaris*), supplied fuel for the bakers' furnaces. Larger trees, such as alders (*Alnus incana*), poplars (*Populus nigra*), and willows (*Salix alba*) growing on some preserved river islands, were cut every seven years for use as firewood and timber. Because these products were equitably distributed among local families, the privatization of the commons that occurred following the French Revolution increased the misery of the poorest.

Reclamation of the Riparian Wetlands

Flash floods occurred frequently in the Rhône-Alpes region during the eighteenth and nineteenth centuries, often destroying crops. Because rivers there flowed from high elevations over deposited alluvium and could easily flood lower elevations, wetlands eventually replaced cultivated fields in many areas.[5] In order to manage the flows of water, sediment, and nutrients over space and time, hydraulic engineers undertook major projects on the Alpine floodplains during the nineteenth century. Much contemporary research focused on controlling floods and erosion, creating shipping channels, and using river sediment as fertilizer.[6] By 1830 population increases in Alpine regions had already led to the deforestation and cultivation of hill slopes; to the displacement downstream of huge volumes of water and sediments, causing instability in the floodplains; and to diking and agricultural development by way of drainage and reclamation processes.

At this time Italian engineers were considered authorities in the art of channelizing rivers to protect crops, diverting their flows into marshes, and using sedimentation processes to create arable soils. These experts recommended that the river be diked and straightened and that wetlands be replaced with mesic habitats. Instead of causing disease and destroying crops, rivers would then become useful for navigation, for irrigation, for depositing fertile alluvial sediment, and as sources of hydraulic energy.

In 1772 King Charles Emmanuel III had ordered the mapping of the Isère River valley in order to initiate a more efficient scheme. Fifteen years later his son and successor, King Victor Amadeus III, ordered a diking study now known as the "Diguement de l'Isère et de l'Arc." Between 1786 and 1789, an engineer named Francesco-Luigi Garella updated and augmented information recorded in 1773 and created a map to accompany the study. Composed of twelve sheets and drawn to a scale of 1/2400, this document depicts the land cover of the valley and the course of the proposed channel.[7] Unfortunately, the study was interrupted by the outbreak of the French Revolution.

During the first annexation of Savoy by France between 1792 and 1814, French administrators found Garella's proposal for diking the Isère interesting and potentially valuable. At least one official of the French government, Paris physician Antoine François, comte de Fourcroy (1755–1809), demanded that riparian wetlands be drained and converted to agricultural use as a public health measure, citing as justification the fact that life expectancy was only twenty years in the vicinity of marshes versus thirty-five years on the more elevated terraces.[8] The role of wetlands in the spread of diseases such as marsh fever had been observed even by the ancient Greeks and Romans, and nineteenth-century engineers likewise viewed marshes and ponds as a threat to public health.[9] Two laws, passed in November 1792 and September 1807, authorized the state not only to carry out drainage projects but also to grant concessions for the execution of such projects to private companies, which would then receive title to significant portions of the reclaimed areas.[10] But before work could get under way, the French occupation ended and Savoy again became part of the kingdom of Sardinia.

A major flood at the end of July 1816 underscored the importance of constructing dikes to protect cultivated areas and to safeguard public health.[11] In the aftermath of French occupation, Savoy's economy remained stagnant and many inhabitants migrated to France. King Victor Emmanuel, seeking a way to unify regions that had been separated for many years, saw the diking scheme as a way to fulfill several objectives. First, it would deter immigration by providing employment for local residents. Second, it would help to prevent soil erosion on both private and public lands. Third, it would allow reclamation of large pastures and marshy fens for production of crops such as cereals, vegetables, hemp, and hay. Fourth, by eliminating marshy habitat, the construction of dikes would reduce the incidence of marsh fever in the area (as Horace Bénédict de Saussure had proposed in *Voyages dans les Alpes* [1796]). And, along with other members of the restored Sardinian government, perhaps Victor Emmanuel hoped that straightening the river between two dikes and concentrating water flows into a single deep channel would establish a firm border between Savoy and France.[12] Because the "Diguement" could help to accomplish all of these objectives, it again came under review on January 7, 1823.

From August 17, 1824, when Garella's proposal received royal approval, to 1828, when the contract for construction of the first section was awarded to a civil engineering company, technical questions concerning the course and shape of the new channel were hotly debated by Sardinian engineers. They finally agreed to construct a pair of permanent dikes made of local alluvial deposits, including fine sand, gravel, and pebbles, and built to a height of 2.4 meters upstream and 3 meters downstream from the Arc River junction in order to contain hundred-year floods. The distance between these dikes was calculated at between 112 and 132 meters. The interior slopes were to be protected from erosion by rocks extracted from the nearby calcareous hills (figure 6.3).[13]

Diking was considered an efficient but expensive solution because it straightened the river into a long artificial canal bounded by continuous dikes high enough to contain it even during high-flow periods and occasional floods. The channel was built in sections by first digging into the sediment and then building levees with the extracted material.[14] Funding problems required the state to subsidize communes or private landowners who were unable to pay the taxes with which the works were to be financed. As a result, large tracts of public land were created along the valley. In October 1854, when the project was completed, the Isère ran straight between two forty-kilometer-long dikes that protected an area of six thousand hectares.

On one hand, diking provided employment for a large number of men, women, and children for nearly thirty years. It also allowed for the protection of some cultivated areas while promoting construction of bridges and, on the embankments, modern transportation arteries such as the Royal Road and the Victor Emmanuel Railway. On the other hand, diking failed to achieve two of its main goals: the expansion of agricultural development and the eradication of marsh fevers. The former floodplain and the old braided channels still flooded frequently during the high-flow period from April to July,

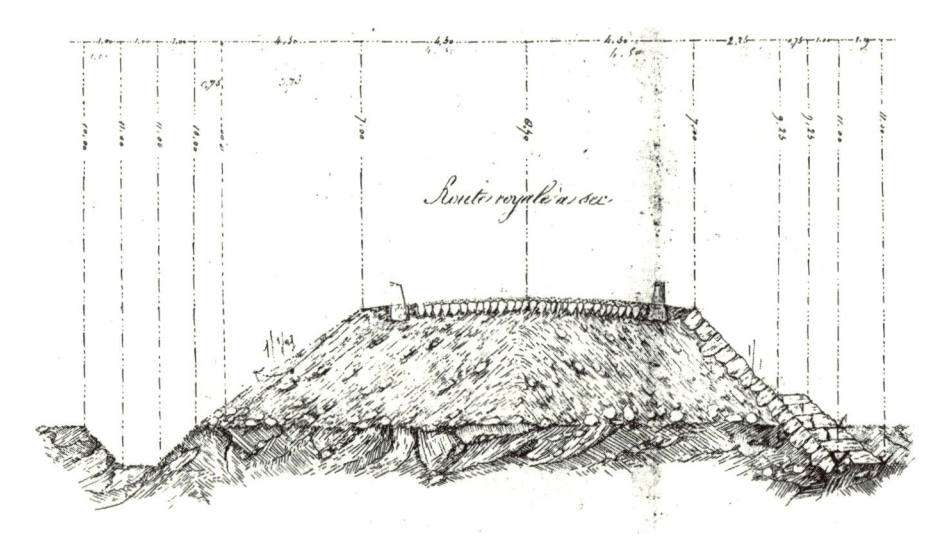

Figure 6.3. A design for the Isère River dikes from Gioacchino Chianale's "Epures de différentes natures de digues" (Archives Départementales de la Savoie, Chambéry, série FS, additions, travaux publics, no. 27 [Chambéry, 1828]).

making cultivation impossible, and a spike in the number of malaria cases was traceable to the mosquito populations that flourished after the annual floods.[15]

To deal with these problems, the Sardinian engineers adopted a technique employed since the sixteenth century in Tuscany's Val di Chiana and called *colmata*.[16] During the nineteenth century this word was rendered in French as the technical term *colmatage,* in German as either *Kolmatierung* or *Aufschlämmung,* and in English as *warping*.[17] The procedure consisted of diverting river flows into marshes and other uncultivated areas, where the sediments carried by the water, such as fine sand, silt, clay, and organic matter, would collect and thus improve the soil for cultivation.[18] Silty water was captured upstream by means of sluices and directed via a carrier canal into large warping basins (figure 6.4). Water flow velocity was slowed significantly in the basins by vegetation and transversal artificial levees so that the sediment load progressively shifted from upstream to downstream; the filtered water was then funneled into a discharge canal and returned to the river downstream. The system became more efficient after two years owing to the growth of plant communities that trapped sediments in the basins.[19]

Development of the Combe de Savoie after 1860

When Savoy became part of France on June 15, 1860, French laws replaced the Sardinian ones, and the engineers of the Ponts et Chausées corps took over the task of repairing the enormous damage from the hundred-year flood of November 1, 1859.[20] A little over a year after these floods, on December 19, 1860, while repairs were still in progress, an organization called the Syndicat de l'Isère et de l'Arc was created to manage dikes and to improve defensive structures. In 1864 the syndicate would play an important role in persuading landowners to warp and drain wetlands for both public health and economic reasons. Many landowners who could not pay the substantial "drainage taxes" were obliged to sell their property to the state.

The French engineers made not only repairs but also improvements: dikes were built higher and additional defense works were developed until 1880.[21] For example, after 1868, Louis Choron used vegetation to prevent erosion caused by flooding.[22] Trees were planted on the exterior slopes of dikes, and

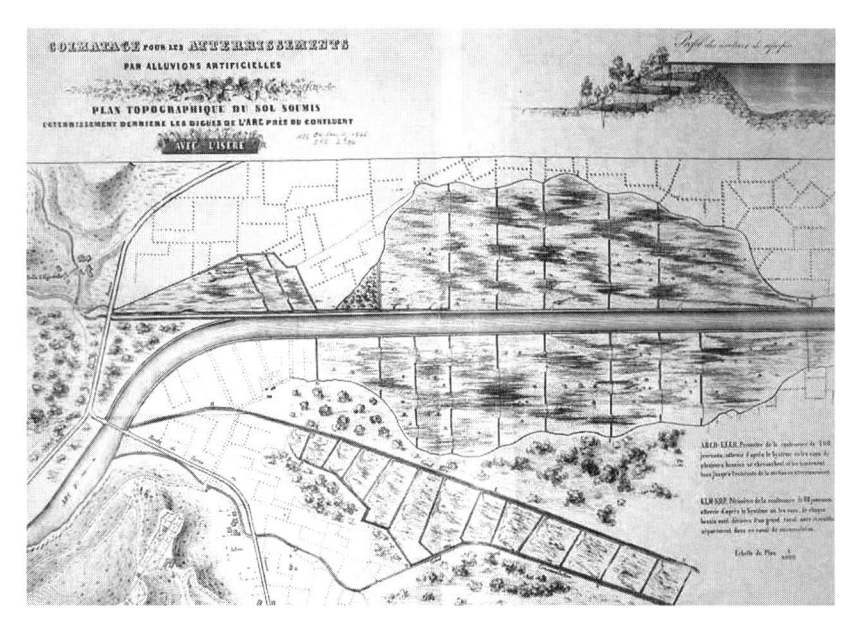

Figure 6.4. "Colmatage pour les Atterrissements par Alluvions Artificielles," a topographical plan of the warping system used in the Combe de Savoie, reproduced from Antoine Chiron's *Atterrissements artificiels dans la vallée de l'Isère, cause des fièvres endémiques et périodiques de cette vallée et moyens de les prévenir* (1846).

grasslands were cultivated on top of them. During this period the dikes were managed by farmers who paid the state for the right to gather wood and harvest hay on them. Warping works spread over protected areas located just outside the dikes. The Isère valley was divided into fifteen warping sections according to water supply, duration of warping operation, and other characteristics. French engineers noted that the Sardinian technique of successive discharges from one basin into the next produced heterogeneous soils, meaning that, from downstream to upstream, basins were characterized by deeper and deeper soils, the grain size of which varied from clay to sand. In order to control heterogeneity of soils over the reclaimed floodplain, French engineers established the simultaneous feeding system.[23] In this procedure similar flows (in terms of volume and sediment content) filled each warping basin at the same rate with silty/sandy soils. This system did not entirely replace the successive discharges system. Each section thus had its own characteristics, depending on which system was used.

Before they could reclaim the basins for agricultural uses, French engineers had to control the groundwater table. The English technique called "underdraining" began to be used in France during the Second Empire.[24] Because water seeped through the dikes during high-flow periods and caused flooding, diking had to be complemented by drainage. This operation improved soil texture and helped to eliminate sources of malaria. A law established on June 21, 1865, required private and communal landowners to cooperate with drainage and reclamation projects recommended by the Ponts et Chaussées engineers.[25] The *associations syndicales* created to implement this new law were charged with harnessing rivers, irrigating farmlands, filling in wetlands, reclaiming unproductive derelict zones, and draining marshes. They also promoted the establishment of large-scale enterprises such as regional drainage networks.[26]

Beginning in 1875, the insect pest phylloxera destroyed many vineyards in the Isère valley, forcing farmers to migrate into the floodplain in order to develop alternative crops, such as potatoes and cereals. About ten syndicates formed there between 1865 and 1900 in order to build and maintain drainage networks for the warping basins that still flooded periodically.[27] As a result, by 1890 many of these basins had become suitable for cultivation. Although some wetlands remained difficult to drain, they continued to be valuable sources of *blache*, which was still widely used to fertilize both plowed fields and the vineyards restored after the phylloxera crisis. Fens and marshland played a major role in the local agricultural economy up to World War I and were particularly prized by farmers, who generally opposed drainage projects.

A statutory order of October 30, 1935, reinforced and extended the law of June 1865. It aimed at improving valley soils by means of extensive drainage networks built at the state's expense. Two years after this order, in 1937, work began on a drainage project along the Isère River valley from Albertville to Grenoble that would not be completed until 1945. Compared with the development of diking and warping systems, accomplished in both cases within a few decades, drainage projects took much longer to complete.[28] The delay was due in part to resistance from area farmers, who used marshes as wet meadows, pastures, and sources of valuable sedge grass through World War II.

After 1945 the effects of rural migration and the widespread use of chemical fertilizers caused interest in marshlands to dwindle. The management of wetlands also changed at this time: *blache* was harvested only in exceptionally dry years, and as it became less useful for agriculture, many marshes were drained and plowed, and some were even transformed into poplar plantations. Beginning in 1970 the popularity of maize as a commercial crop led to the drainage and subsequent cultivation of many wetlands, and the remaining marshy areas progressively evolved to dense thickets of willows and alders. This evolution had two impacts: a significant decrease in biodiversity linked to the reduction of habitats, which also corresponded to diminished interest in hunting and in the aesthetic value of landscape; and large increases in mosquito populations due to the neglect of drainage networks.

At present, only the former warping basins located below the level of the river still sustain wet plant communities. These areas, containing soils made of alluvial gravel deposits covered by thin layers of silty clay, are characterized by a water table that varies from −0.20 meters to −0.50 meters over the course of the year. Since 1990 they have been protected as wetland habitats.

THE HYDRO-ENGINEERING projects that determined the present-day landscape of the Alpine valleys were largely executed during the nineteenth century. Two major periods of engineering works can be distinguished. From the French Revolution to 1830, physiocrats and agronomists wanted to convert wetlands into "useful" properties.[29] After 1830 authorities undertook diking projects for a broader range of socioeconomic and political reasons—that is, not only to increase the amount of arable land and serve the public health by combating marsh fever but also to discourage emigration and reestablish a Sardinian national identity in the aftermath of French occupation. The role of large-scale public works in unifying multilingual and multicultural Alpine regions such as Savoy and Switzerland should not be underestimated. The Isère River diking scheme succeeded in uniting a heterogeneous group of stakeholders from disparate social and cultural backgrounds around the common goal of greater economic stability for the region. Thus, like the hydro-engineering scheme of Switzerland's Linth Valley, the "Diguement de l'Isère" symbolized the strength of a cohesive human community versus nature.[30]

Today, evidence of this Herculean enterprise remains visible in the straightened channels flowing between two dikes and in the engineered landscapes produced by a complex history of artificial silting and drainage. Most of the Alpine floodplains appear as mosaics of fields, fens, and meadows whose soils are the result of recent anthropogenic activities.[31] Many areas are under cultivation, while many others are still covered by "wild" vegetation that was originally planted and maintained by humans.

7

Holding the Line

Pollution, Power, and Rivers in Yorkshire and the Ruhr, 1850–1990

CHARLES E. CLOSMANN

For much of the nineteenth and early twentieth centuries the process of supplying water to European cities entailed massive environmental costs. Urban centers consumed huge volumes of water and returned it as sewage to nearby rivers, where it often created an odorous nuisance and a threat to public health for downstream residents. Industry also used vast quantities of water for cooling and other processes before discharging effluent into the nearest river, lake, or estuary. The challenge for governments in heavily industrialized areas was to guarantee enough fresh water to protect public health and economic growth without causing irreparable harm to local rivers or streams.[1]

In a series of essays from the 1990s, historical geographer John Sheail challenges the view that a strong central government is best suited to create a sustainable balance between these two objectives. Focusing mainly on industrialized Yorkshire between 1850 and 1960, Sheail argues that England's decentralized political framework fostered a workable trade-off between those interests demanding a plentiful supply of clean water and those wanting to limit stream pollution. The English approach also contrasted favorably with that of the German empire, according to Sheail. Citing the pioneering work of historian Franz-Josef Brüggemeier, Sheail notes that in the early 1900s Prussia's authoritarian centralized government facilitated the transformation of the Ruhr region into an "industrial preserve" where extreme ecological damage was more or less deliberately concentrated in a few rivers in order to protect the quality of streams elsewhere in Germany. The local population—

mostly working-class migrants to the region—barely resisted, so dependent were they on nearby mines and factories for their jobs.[2] Reflecting on England's relative superiority over Prussia in the management of waterways, Sheail asserts that "a world of difference exists between having authority and using it to the greatest possible effect."[3]

Sheail's conclusions about government's ability to sustainably manage water resources invite further scrutiny. I test Sheail's argument by comparing water management policies in Yorkshire and the Ruhr from the beginning of the Industrial Revolution to the end of the Cold War. This comparison makes sense for three reasons. In both regions coal mining and heavy industry dominated the local economies into the 1960s before beginning a thirty-year decline. Moreover, both regions witnessed a startling increase in the quantity and variety of industrial chemicals discharged into local streams after World War II. Finally, the government agencies and legal frameworks created to manage water in each country differed in particular ways for much of this period.[4] Accordingly, such a comparison highlights the effectiveness of contrasting approaches in the management of heavily industrialized waterways.

The impact of government structures on the development of sustainable water policies diminished after 1970. In fact, two separate issues were critical to balancing growth, jobs, and acceptable levels of river pollution. At the most basic level the emergence of modern environmentalism in the late 1960s encouraged governments in both countries to impose tighter restrictions on water discharges by all users. Second, the relative financial strengths of Britain and Germany during the postwar years outweighed even the best-organized, most well-intentioned policies to curtail pollution. Hence, the twenty years after 1970 were themselves a watershed in government's ability to manage water in a sustainable way.

England's Yorkshire Region

Beginning in the mid-nineteenth century, Yorkshire emerged as a center of English industrial production. With low mountains in the west and high moorlands in the northeast, the region is drained by numerous rivers flowing west to east, all of which eventually join the Humber directly or flow into tributaries of the Humber, such as the Ouse or Trent (figure 7.1). Most of these streams are quite short—the Humber is only about forty miles long, while the Ouse, Aire, Calder, Nidd, and Wharfe are all under a hundred miles long—but some are navigable for much of their length, a factor that encouraged industrial growth in this region during the nineteenth century.

The vast seams of coal lying near the surface of the land were also critical to Yorkshire's industrial development. Especially in the area known as the West Riding of Yorkshire, hundreds of textile mills and factories emerged in the late eighteenth and early nineteenth centuries to take advantage of favorable transportation routes, mineral wealth, and a large work force. Towns like Leeds, Bradford, and Sheffield grew into big cities during the middle and later years of the nineteenth century. In Leeds the textile mills drew thousands of workers from the countryside, and the population exploded from about 53,000 in 1800 to 400,000 by 1990. Other cities grew at comparable rates.[5]

At the same time, Yorkshire's cities, mines, and factories produced millions of gallons of wastewater, much of which was discharged directly into the Aire, Calder, Wharfe, and other streams. Downstream from the wastewater

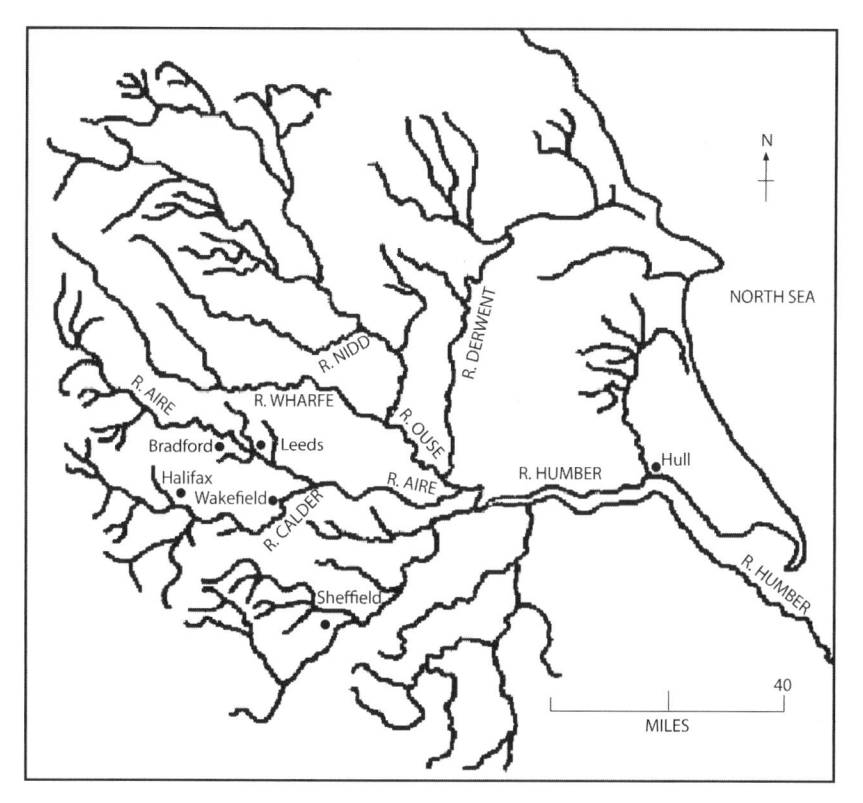

Figure 7.1. Map of rivers in Yorkshire, a county in the north of England. Created by Donna Durham and reproduced by permission.

outfalls, reeking sewage and trade waste imperiled municipal water supplies, stocks of fish, and those tanneries, textile mills, and other firms that depended on plentiful freshwater for cooling, cleaning, and other purposes. Commenting years later, the noted sanitary expert William P. Dunbar vividly described how accumulating waste had transformed Yorkshire's small streams into "ill smelling ... inky colored" waterways, choked with sludge from the cumulative excrement of some 3 million people and the effluent from hundreds of factories, mines, and other concerns.[6]

English Water Policy, 1850–1918

English political and legal customs posed a considerable challenge in the effort to limit riparian pollution while supplying ample freshwater for both residential and industrial uses. Most importantly, a tradition of decentralized government and the fact that a variety of towns, cities, and counties shared with Parliament jurisdiction over the rivers meant that progress toward this goal was halting at best. Nevertheless, reformers grappled with this issue in typically Victorian fashion, establishing the Royal Sewage Commission to study pollution, a problem they described in 1857 as "an evil of national importance."[7] The commission recommended, somewhat unhelpfully, that towns causing the contamination of local rivers should stop doing so.[8] Ten years later the deplorable state of the Aire, Calder, and other streams prompted the commission to call for more effective remedies than those provided by existing riparian law.[9] Under common law all landowners had the right to use the water flowing past their property as they saw fit and to expect it to be in a natural state, subject to "reasonable" use by upstream interests. Yet English common law proved unable to deal with the expanding use of sewers and the discharge of trade effluents into local streams. This prompted the commission to call for a more centralized statutory policy for all of England and Wales.[10]

While ambitious, the legislation adopted by Parliament failed to create the conditions necessary to achieve major, systematic improvements to water quality in Yorkshire or elsewhere. The Rivers (Prevention of Pollution) Act of 1876, for instance, allowed traders to discharge effluent into local streams so long as they claimed they had taken the "best practicable ... means to render harmless the poisonous, noxious or polluting liquid so falling, flowing or carried into the streams."[11] Under an earlier public health act of 1875, factories could legally dump wastes into public sewers and treatment works subject to some restrictions. Yet the same legislation permitted a local sanitary authority to refuse to accept wastes if the sewers were too small, if industrial wastes might harm the sewer, or if those wastes threatened public health. Ironically, the combined effect of the two acts was often to prevent factories from using treatment works at all and to encourage them to pour their untreated effluent directly into local streams or soakage pits.[12] Finally, the Rivers Act of 1876 —which remained in effect until 1951—maintained England's tradition of decentralized administration, creating a framework in which responsibility for water supply and disposal rested with a jumble of interests, many of them determined to protect private property. Referring to the act of 1876, Sheail states that, "instead of a unity of purpose, it reflected many agendas, each addressing its particular type of locality."[13] Other legislation, including a local government act of 1888, only bolstered tension between riparian stakeholders by dividing the country into several administrative counties (including the West Riding of Yorkshire County Council) and adding yet another layer of bureaucracy.[14]

Yet, as Sheail contends, this confusing political framework actually set the stage for a rough, though far from perfect, balance between economic growth and the control of river pollution during the late 1800s and early 1900s. Among other things, this system worked because it allowed stakeholders access to the political system at various points and because it brought to light the rights and arguments of aggrieved parties for all to see. Two cases demonstrate how this process operated. In a notable example from 1887, the largest estate owner along the Calder successfully petitioned Parliament to force the upstream city of Halifax to return a higher percentage of relatively clean "compensation water" to the Calder, thus reducing contamination downstream. Essentially, the House of Lords acted to protect the property rights of a major landowner. Complainants might also adopt a somewhat less confrontational approach. Faced with huge volumes of waste flowing from Bradford's sewer outfalls, the West Riding of Yorkshire County Council joined with the city to condemn the seventeen-hundred-acre Estolt Park in order to construct a more effective combined sewage plant. After many protests by the landowners, the House of Commons intervened in 1901 and condemned only three hundred acres of the estate—a clear case of Parliament favoring private property while also recognizing that pollution near Bradford had reached catastrophic proportions. In both examples diverse interests staked out their positions for all to see, allowing Parliament to step in and arbitrate an imperfect but pragmatic solution.[15]

Businesses often played a critical role in local sanitary improvements. In Wakefield, for instance, a private company established in 1837 supplied water to the town from a location two miles downstream on the River Calder. Yet by the 1870s the situation became dire for Wakefield residents and owners of the nearby woolen mills: sewage from hundreds of thousands of residents along the Calder flowed past the waterworks each day and tainted Wakefield's water supply. In response, the water company petitioned the House of Commons for permission to transport pure, hygienic water from wells near Wath-on-Deane, some twenty miles distant. While this solution may have reduced waterborne diseases in Wakefield, it would also have increased the level of minerals in Wakefield's water, making it "hard" and unsuitable for use by the Henry and Lee woolen mills. Standing in support of the wool trade, Wakefield's council members opposed the plan for a private water supply near Wath, and a long conflict began. In another example of hardnosed calculations, Wakefield eventually paid the sum of £210,000 to take over the water company and construct a new waterworks that would better serve the town and its employers. While the cost was high, it reflected the city's desire to protect both public health and the local wool trade.[16]

Scientists often lined up on both sides of these debates, sharpening the sense of conflict between stakeholders. In the Wakefield case, for example, the noted chemist Edwin Frankland took the side of Wakefield's private water company, contrasting the health benefits of a salubrious water supply near Wath-on-Deane with the vile nature of the town's existing drinking water. In contrast, city experts cited the high mineral content in water near Wath. Yet all parties knew the stakes involved and the position of the wool trade, a factor that prolonged but did not prevent a practical resolution of the issues at hand.[17] According to Christopher Hamlin, interested parties often translated scientific concerns about pollution into commonsense questions of public health, civic pride, and sustained prosperity. The *Wakefield Express,* for instance, argued that an inadequate water supply would prevent the town from "ever attaining that position to which its central situation, railway facilities, and its almost unsurpassed natural setting entitles it."[18] It was an approach to public welfare that succeeded as much from conflict and parochial self-interest as from any broader vision of the common good.

English Water Policy, 1919–1970

The interwar years put England's highly decentralized system of water management to the test, forcing it to evolve. Beginning in the early 1930s, the mounting demand for plentiful clean water outpaced the resources of government agencies, and experts warned that water shortages threatened economic growth, jobs, agriculture, and public health.[19] Focusing more explicitly on water quality, spokespersons from *Salmon and Trout Magazine,* the British Field Sports Society, and other recreational groups complained bitterly about the perilous state of some rivers.[20] Capturing these concerns in 1937, Labour Party MP Alfred Short described the River Don, near Sheffield, as "flowing in agony, ashamed of itself."[21]

The system of water management also began to collapse under the weight of its own complexity. As R. C. S. Walters noted in 1947, "Legislation with regard to water rights in Britain has grown up mostly as the result of circumstances. It is haphazard and has been carried out piecemeal." He went on to say, "Recently, it has been felt that the legal position of public water supplies in Britain is unsatisfactory in many ways, and it is probable that in the future the law will be simplified."[22] Moreover, few of the existing river agencies enjoyed authority for an entire watershed, and none controlled all aspects of water management in their jurisdictions.[23]

The problem was acute in Yorkshire, as elsewhere. In the West Riding of Yorkshire the county council established the River Ouse Catchment Board to manage issues of drainage. This encouraged Parliament to pass the Land Drainage Act of 1930, which authorized the creation of even more drainage boards throughout Britain. In addition, the Yorkshire Fisheries District regulated fishing along the county's streams.[24] Such authorities had powers beyond those already granted to Bradford, Leeds, and Sheffield; to the West Riding of Yorkshire County Council; and to the West Riding of Yorkshire Rivers Board. Faced with a severe water shortage during a drought in 1933, Parliament also created the West Riding of Yorkshire Regional Advisory Commission, which encompassed an astounding 142 individual authorities responsible for water use and disposal.[25]

As in the late nineteenth century, public pressure prompted changes in the English system of water management, especially in the years just after World War II. In this case, demands for change came from technical experts, from a general public that increasingly turned to government for solutions to many problems, and from a variety of other interests. In an article for the trade publication *Water & Sewage Works* in 1949, Paul Smith captured the spirit of the times, stating, "It has long been recognized that centralized control of Britain's water supplies is an urgent need. . . . New legislation will remove planning from the confined, parochial sphere to the Central Government."[26]

Acting on this mandate, Parliament passed measures to consolidate England's river entities and to tighten restrictions on wastewater releases into English streams. According to the River Boards Act of 1948, individual river boards could now manage issues of fishing, flood control, and pollution within an entire watershed. The act of 1948 also made the new river boards large enough to hire technical experts capable of surveying, testing, and analyzing water quality.[27] The two Prevention of Pollution acts, passed in 1951 and 1961, went further, attaching minimum standards of treatment for all waste discharges into English streams, depending on the locality.[28] New laws also built upon the Drainage of Trade Premises Act of 1937. A key step in solving the problem of untreated industrial waste, the act encouraged factories to pour their effluent into public sewers and treatment plants rather than directly into rivers, subject to certain financial charges or pretreatment requirements.[29] Finally, the Water Resources Act of 1963 established twenty-nine river authorities in England and Wales, empowering them to monitor the water supplies, drainage, pollution, and fisheries within an entire region. With a majority of members appointed by local industry, waterworks, sewage authorities, and fisheries, each river authority was "representative of all river users" and capable of reaching solutions that were "fair and usually acceptable" to the public, according to one expert. In Yorkshire, the Yorkshire Ouse and Hull River Authority fulfilled these duties for several rivers flowing into the Humber and its estuary.[30]

Over time this more comprehensive framework of water management policies began to show results. As early as the mid-1950s, public sewage treatment plants used the Drainage of Trade Premises Act of 1937 to limit damage to waterways by acids, pickling liquors, and other industrial by-products.[31] The act proved especially successful in heavily industrialized Yorkshire, at least in the fifteen years following its passage. In 1940, J. H. Garner, chief inspector of the West Riding of Yorkshire River Board, reported that the percentage of factories discharging their waste in sewers had doubled since the late nineteenth century. Twelve years later an inspector from the Ministry of Agriculture and Fisheries described the measure as a "very great technical advance," in part because local councils encouraged industry and sewer districts to cooperate.[32] The framework for water management also succeeded because it encouraged a "striving for balance," according to Sheail. He notes, for instance, that the new river boards could achieve a more coherent approach to the management of streams, primarily because they brought stakeholders under one umbrella but also because they demonstrated concern for the various constituencies affected by water and sewage policies.[33]

As with any history of environmental conditions, this success story must be qualified. After World War II, per capita water consumption skyrocketed as the British economy grew at a steady rate of 2.5 percent and as households demanded such amenities as bathtubs, fixed basins, and other types of plumbing. Moreover, the burgeoning electricity, paper, and chemical industries demanded higher amounts of water for their own processes, and England, a country rich in natural sources of water, began to experience water shortages.[34] These huge amounts of industrial wastewater often overtaxed public sewer systems, especially in the late 1950s. Commenting in 1957, sanitation expert John Finch noted that industrial wastes were causing "sizeable headaches for many local authorities" and that "many are simply overwhelmed by a problem which becomes more acute with the passing of years."[35]

Nevertheless, by the late 1960s water quality had improved along many streams. In 1969, for instance, the chief purification officer of the Thames Conservancy stated that "the record of the river authorities in pollution prevention is a good one: not only has a vast amount of new pollution been prevented, but much long-standing pollution has been eliminated."[36] Except for the exact timing of major improvements, a Department of the Environment survey in 1975 concurred, noting steady reductions in the number of "Category III" (poor quality) and "Category IV" (grossly polluted) streams during the previous four years. Without directly attributing improvements in water quality to increased spending on sewage treatment, the government's expert also maintained that large amounts of money were spent annually in England and Wales to meet national water quality standards for major wastewater discharges.[37]

English Water Policy to 1990

After 1970 the post–World War II trend toward consolidating functions within an entire watershed continued. The lynchpin of this approach, the Water Act of 1973, combined all of England and Wales's twenty-nine river authorities and more than a thousand waterworks, sewage authorities, and other entities into ten large regional water authorities (RWAs). For the first time, regional organizations had responsibility for the whole water cycle, from supply to disposal to pollution control. In Yorkshire, for instance, the Yorkshire Water Authority now supervised water supply and disposal within the Humber catchment. Moreover, Parliament nationalized ownership of RWAs while making them responsible for financing their own operations.[38] Finally, the new entities maintained the pluralistic structure of previous river boards, giving local representatives majority membership in the RWAs.[39]

While impressive, this new framework for water management faced critical challenges in the 1970s and 1980s. For a variety of reasons, national policy in Great Britain increasingly emphasized water *quality* and river pollution rather than the more traditional emphasis on providing plentiful *quantities* of water for the nation's diverse users. In part, England's water authorities were the victim of their own past achievements. One expert, M. B. Beck of the Institute of Statisticians, explained it historically, noting that England's water management system had effectively dealt with "pathogenic bacterial" threats to public health in the nineteenth century and with "gross pollution" caused by suspended solids (SS) and biochemical oxygen demand (BOD) in the period up to about 1970. Having controlled these forms of pollution, the RWAs were now asked by the national government and the European Economic Community (EEC) to meet higher standards focused on long-term human exposure to chemical wastes, a growing concern since the end of World War II.[40] Undoubtedly, the emergence of the environmental movement in Great Britain and other parts of Europe also caused a shift in British and European policy, and focused more attention on water quality and pollution. In 1970, for instance, Parliament established the Department of the Environment and charged it with setting national water priorities. This initiative led to the transfer of the national Water Pollution Research Lab to its own jurisdiction.[41]

Eventually, financial problems posed an even greater challenge to agencies like the Yorkshire Water Authority.[42] Beginning in the mid-1970s, chronic underinvestment plagued the RWAs, especially during the years of Prime Minister Margaret Thatcher's government (1979–1990). Adhering to a strict "monetarist" policy, Thatcher's Conservative government attempted to reduce public debt, in part by limiting the RWAs' ability to borrow money, a tactic that effectively postponed necessary upgrades of sewage treatment plants.[43] This "scissor effect," characterized by rising water quality standards and shrinking investment, led to a steady degradation in rivers and to England's reputation as the "dirty man of Europe."[44]

England began to reclaim its reputation for protecting waterways only in the 1990s. In 1989 the Thatcher government sold off the RWAs for about £28 billion, privatizing the entire water cycle on the grounds that private companies were more efficient than government agencies and more capable of raising money on the capital markets. As a consequence of these changes, Yorkshire Water Services emerged to serve users in the Humber estuary region. The results—in general—have been encouraging. Under a continued regimen of government regulation and EU mandates, water quality along most English and Welsh rivers has once again improved, and England seems to have shed the nasty reputation it earned during the 1970s and 1980s.[45]

Germany's Ruhr Region

Geographically, the Ruhr region of Germany is strikingly similar to Yorkshire. Part of a wide plain extending from northern France to northwestern Germany, the Ruhr is actually a small area (200 square miles) just east of the Rhine River. Of the numerous streams that drain this landscape of low hills, broad valleys, and high rainfall, the most important are the Ruhr, Emscher, and Lippe rivers. All are relatively short, and all meander east to west, emptying into the Rhine near Düsseldorf (figure 7.2). During the nineteenth century these waterways were generally unsuited to transport. At the same time, however, the region was well situated for exporting products because of its proximity to the Rhine, Europe's busiest trade artery in the 1800s. Finally, the region's mineral wealth became a crucial factor in the Ruhr's startling economic growth in the late nineteenth century.[46]

The presence of large deposits of bituminous coal set the stage for an industrial revolution that began gradually and then rapidly outpaced that of Britain. Coal was mined in the Ruhr at an escalating rate from the middle of the eighteenth century until the 1870s. In 1879 a major technical breakthrough in steel manufacturing—the Gilchrist-Thomas method—rather suddenly made German coal more desirable for the production of high-grade steel. Spurred on by growing markets for manufactured goods, Ruhr mining inter-

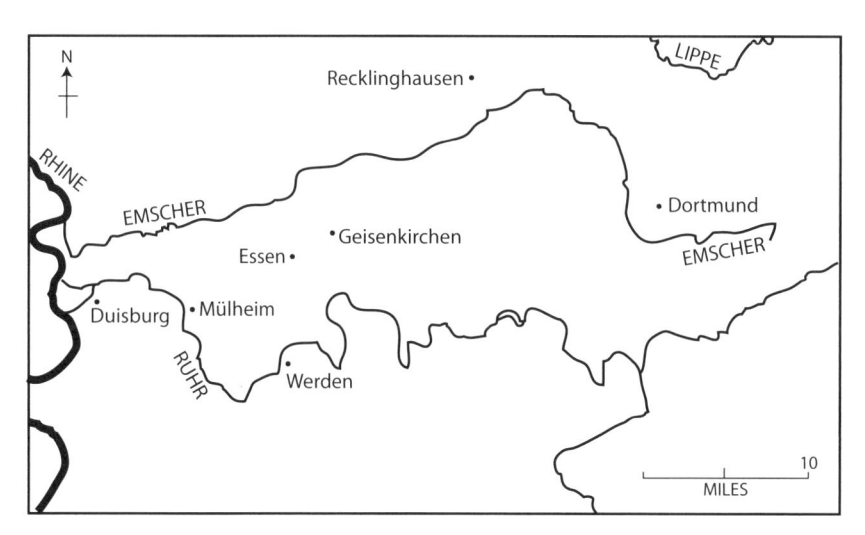

Figure 7.2. Map of waterways in the Ruhr region of Germany. Created by Donna Durham and reproduced by permission.

ests extracted more than 100 million tons of coal annually by 1913, an astounding 1,600-fold increase over the year 1871. Ruhr steel mills also profited from the availability of so much coal: by the 1880s the Rheinische Stahlwerke, Phönix, and Krupp concerns were casting more than half of the world's steel.[47] Attracting thousands of young migrants, towns in the Ruhr expanded quickly and in a highly chaotic manner. Industrialization transformed Essen, a small town of only 4,000 inhabitants in 1800, to a bustling city of more than 400,000 by the year 1900, and once-sleepy towns like Dortmund, Duisburg, and Gelsenkirchen grew at comparable rates.[48]

German Water Policy, 1850–1945

Superficially, the legal and governmental structures established to manage water in the Ruhr were similar to those of Yorkshire. In late nineteenth-century Germany, civil law allowed landowners to petition courts for redress if their property was harmed by the actions of another, while a commercial ordinance of 1871 permitted property holders along rivers to complain if polluted water caused "considerable disadvantages, dangers, or nuisances" to their land.[49] Reflecting centuries of fragmented rule, the Bismarckian constitution assigned responsibility for rivers to the individual states, although much of this (state) power gradually devolved to individual communities.[50] As in Britain, confident bourgeois reformers expanded their influence over city government during the Industrial Revolution, enlarging the role of cities in overseeing such basics as water supply, sewerage, and hygiene. Public health experts, sanitation engineers, and city planners were especially active in this respect, forming the German Society for Public Health in the 1870s and pressuring cities to adopt sewer systems and other amenities.[51] The Ruhr was similar to Yorkshire in other respects as well. In the Emscher valley 2 Prussian provinces (Rhineland and Westphalia), 3 governing districts, 6 cities, 43 departments, and 137 counties held sway, making a unified water policy exceedingly difficult to achieve.[52]

At the same time, however, the overwhelming influence of Prussia within the German Empire shaped the way that local entities articulated their own power, especially in the Ruhr. Awarded the Rhineland in 1815, the authoritarian state of Prussia then controlled the Ruhr, its mineral wealth, and the rivers that drained the region. As Mark Cioc notes, the Ruhr's enormous deposits of coal fueled Prussian militarism and Germany's emergence as Europe's leading producer of coal, iron, and steel.[53] Moreover, officials in the Prussian capital of Berlin were determined to protect and develop the coal and steel industries at almost any cost. Consequently they granted control of the water supply and waste disposal operations to the region's industrial giants.[54]

Beginning in the 1880s severe shortages of clean water threatened to slow the region's extraordinary growth. The mining facilities and the towns that lined the Ruhr River drew their water supply from wells beside the river. At first the water was relatively clean: a five-to-eight-meter-wide bed of sandstone along the banks acted as a natural filter and cleansed the water of most impurities. In the interest of protecting their water supply from contamination, many of the towns, mines, and foundries along the river piped their waste to the Emscher River, which fast became an open-air sewer for the region. By 1900 residents and industries along the Ruhr were consuming hundreds of millions of gallons of water and returning little to the stream itself. This simplistic approach to water management had devastating consequences: in a region with plentiful rainfall, frequent shortages of water became a problem of "towering importance," according to one contemporary observer.[55] Reacting to a perceived crisis, the industrialist August Thyssen warned that catastrophic water shortages could mean unemployment for thousands, while mine directors in the region warned that a scarcity of water for dampening coal dust could lead to deadly explosions in the collieries.[56]

Where industry used the Ruhr to dispose of waste, it also became badly polluted. In 1908 alone businesses discharged 56 million cubic meters of wastewater into the stream, killing fish and fouling the valuable sandbanks with slime.[57] In 1911 the biologist August Thienemann described the Ruhr by Mulheim as a "brown black brew, reeking of prussic acid, containing no trace of oxygen, and absolutely dead."[58]

The Emscher was even filthier than the Ruhr. As stated above, owners of the coal mines treated the Emscher as an open sewer canal, as did Essen and several other cities in the region. By the early 1900s 1.5 million people, 150 mines, and 100 factories poured their effluents into the river, effectively doubling the flow of the slow-moving stream. Indeed, only about 50 percent of the river's total volume came from natural sources; the rest consisted mostly of industrial waste (about 90 percent) and sewage (about 10 percent). Consequently the incidence of waterborne disease was high compared to the rest of Germany; the typhoid infection rate, for instance, was double the average rate for all of Prussia between 1887 and 1900.[59]

Prussia's legal system was completely inadequate to deal with this potentially catastrophic situation. As the water crisis grew, farmers and cities filed dozens of complaints in an attempt to enjoin the worst polluters from discharging effluent into the region's waterways. In 1884, for example, aggrieved parties filed twenty-eight petitions against coal mines near Dortmund for polluting the Emscher, while in the same year farmers complained another fifty times against the city of Essen for discharging sewage with no treatment. Yet many farmers were too poor to proceed against mining interests or cities, and the authorities often refused to challenge large employers. The confusing maze of jurisdictions also hindered a coordinated approach to the problem. The region's industrial interests exploited the chaotic legal system, rarely building treatment plants capable of purifying the massive volumes of waste their own factories produced.[60]

In addition, no river boards existed in the Ruhr to bring the dozens of aggrieved parties together in healthy debate, nor had Germany's parliament, the Reichstag, passed a comprehensive wastewater law for the entire country. Despite mounting calls for such legislation from individual cities, some national politicians, and the German Council for Agriculture, the Reichstag refused to take action, arguing that such matters should be left to the states. Moreover, efforts to pass a comprehensive wastewater law for Prussia languished until 1913, primarily because the representatives of Prussian cities, industry, and agriculture were still disputing the necessity of statewide wastewater standards.[61] When finally adopted, the law asserted that permission for the release of industrial waste would be granted only on the condition that it did not harm the "common good," an extraordinarily subjective standard. In practice, this law—in effect throughout much of Germany until 1957—had only a marginal impact.[62]

Faced with a growing catastrophe, officials from the towns and cities in the Ruhr took matters into their own hands. In 1899 Mayor Weigert of Essen proposed the creation of the Emscher Cooperative to distribute the cost of wastewater treatment among all wastewater-producing entities along the river. According to a draft version of the law creating this cooperative, towns, cities, and businesses in the region were to govern themselves, and towns and cities were to have a deciding vote among all members regarding plans for local wastewater treatment. As the law worked its way through the provincial assemblies of Westphalia and the Rhineland, the mining industry insisted on having a controlling vote in all matters brought before the cooperative's board of directors. Politicians in the provincial and Prussian parliaments agreed, voting the Emscher Cooperative into life in 1904, subject to this major condition. In effect, elected officials abdicated responsibility for Prussia's rivers and ceded control of the Emscher Cooperative to the mining and metallurgical industries, a decision that would have far-reaching consequences.[63]

From the beginning the cooperative promoted wastewater disposal at the lowest possible cost. Guided by this principle, the cooperative immediately began lining the Emscher and its tributaries with concrete and shortening it by almost thirty kilometers in order to give the river a faster flow. Intended in part to reduce flooding along the marshy stream, this giant engineering project also ensured that the river would remain an open-air sewage canal for years to come.[64] Cost minimization also shaped the cooperative's approach to the construction of treatment plants. Protected by its legal status, the cooperative consistently operated at the low end of the technological scale when building purification plants. The cooperative's scientists justified this policy, in part, by arguing that the Rhine—into which the Emscher emptied—could dilute the massive volumes of chloride salts, lime, and other material discharged from the Emscher watershed each day.[65] The cooperative also maintained—as did many scientists at the time—that industrial contaminants posed little threat to human health and that bacteria carried in human waste constituted a much greater hazard.[66]

When the state pressured the cooperative to treat domestic sewage along the Emscher, the cooperative resisted using the latest technology. In one instance, the supervisory authority recommended the construction of twenty-three state-of-the-art biological treatment plants, but the cooperative refused on the grounds that such plants were too expensive. Instead they built a series of 138 "Imhoff tanks," which constituted a cheaper and less effective method of mechanical purification that removed the most noxious materials from raw waste and pumped the remaining water back into the river.[67]

This approach to water pollution changed little over time. As the population of the Emscher valley increased from 1 million to 2.5 million between 1895 and 1925 and mines and steel mills doubled their production, the cooperative continued to employ the cheapest methods of wastewater treatment.[68] To reduce the worst odors associated with raw sewage, the cooperative built nine treatment plants in the early and mid-1920s. Yet more effective biological treatment methods played only a minor role. When politicians described the Emscher as a "river from hell," the cooperative's engineers remained relatively unmoved, opting for less expensive and less effective sewage treatment methods over more advanced technologies.[69]

In the south, the major interests drew their water supply from the Ruhr and adopted an approach similar to that of the Emscher Cooperative. To alleviate the persistent water shortages of the late 1800s, waterworks and power plants in the region created the Ruhr Dam Association (Ruhrtalsperrenverein) in 1899. A consortium of water users, the association constructed a series of reservoirs in the mountains surrounding the Ruhr catchment, a strategy that succeeded in supplying more than 500 million cubic meters of water per year to the region's industries and towns well into the 1930s. Not only did these reservoirs deliver relatively clean water to local users but they also helped to prevent legal battles between industrial firms and waterworks competing for water.[70]

Another cooperative association, the Cooperative for a Clean Ruhr (hereafter the Ruhr Cooperative), was founded in 1911 to protect the Ruhr water supply from pollution. Functioning almost exactly as the Emscher Cooperative did, it quickly moved wastewater to the Rhine through short canals, built a mechanical grate system to reduce oil and tar near the mouth of the river, and constructed surface reservoirs to settle out metallic substances from local factories. Like the Emscher Cooperative, the Ruhr Cooperative faced substantial resistance from commercial fishing and other interests, especially in relation to phenol. A by-product of the coal-tar industry, phenol gave off a strong odor that irritated residents and imparted an unpleasant taste to salmon, eels, and other species of fish taken from the Rhine and other rivers. Because the region used so much water, phenol also entered the water supplies of some communities, including Essen's in 1925. Yet even in the face of widespread criticism the cooperative delayed taking any action, constructing several dephenolization plants only after the issue threatened to become a public relations catastrophe for industry.[71]

The Prussian courts also reinforced the distinctive nature of water management policies in the Ruhr, effectively shielding industry from legal actions. As stated earlier, German commercial laws made it difficult for petitioners to prevent polluting activities, especially if defendants were major employers. In addition, legal rulings evolved to protect activities and levels of pollution deemed normal in a particular locale. A Supreme Court ruling in 1916 declared, for example, that it was illogical for a local farmer to expect his fruit trees to flourish near a coking factory since coking plants were common in the region, while fruit orchards were not. According to Franz-Josef Brüggemeier, this type of argument evolved over time as more industry located in the Ruhr, effectively creating an "industrial preserve" in which mines, factories, and cities were exempt from complaints.[72]

In most cases, residents also offered little opposition to the region's biggest polluters. The largely working-class population took great pride in the smokestacks and factories that dominated the landscape, a region sometimes called Prussia's "Wild West." Noticing the rainbow of purple, brown, and red swirling in the filthy Wupper—an industrialized stream in the south part of the Ruhr—a town mayor pithily observed in 1926 that, "for only so long as the

Wupper is dirty, there is still work to be found."[73] The results of this attitude by the late 1930s were quite predictable, as water in the Ruhr remained usable by industry but unsafe to drink.

German Water Policy, 1946–1970

As in Yorkshire, economic conditions in the twenty years after World War II exacerbated problems of water management. Personal and industrial consumption of water grew rapidly, and a variety of new chemical by-products placed new demands on existing treatment systems. The oil industry in particular demanded huge quantities of water and returned much of it to the rivers of the Federal Republic of Germany (FRG) laden with petroleum waste. Detergents also contaminated local streams, as households grew more prosperous and used more water.[74] By the early 1950s, pollution imperiled the FRG's ability to sustain its "economic miracle" with adequate supplies of freshwater. Warning of another crisis, journalists, scientists, and politicians formed lobbying groups such as the Alliance for the Protection of Germany's Waters, demanded action on the long-delayed national pollution law, and advocated the complete biological purification of all wastewater.[75]

The problem was especially acute in the Ruhr. German politicians considered the Ruhr a lynchpin of the country's economic recovery after the war and therefore a bulwark against communism. Consequently, industrial production, water consumption, and pollution increased steadily from the 1950s and into the 1960s. In particular, the chemical content of household wastewater rose steadily, while some two thousand individual firms discharged heavy-metal tailings into the region's streams. From 1950 to 1969 the level of organic and inorganic material in local waterways rose by 12 percent, as did quantities of harmful bacteria. At a time when water agencies in England were getting river pollution under control, rivers and streams in the Ruhr were deteriorating.[76]

Yet the political and legal institutions with oversight for water management barely evolved to meet these new demands. In contrast to England, where pollution laws changed substantially from the 1940s to the 1960s, those in the FRG hardly changed at all. Consistent with a political tradition that left much power with the states, government officials in the FRG delayed action on passing a national water law until 1957, some sixty years after such legislation was first proposed.[77] In addition, courts in North Rhine–Westphalia continued to reinforce the doctrine that the Ruhr was an "industrial preserve," a region where pollution was the norm. In a court case in Duisburg in

1951, for instance, the judge declared that residents would have to adapt to smoke belching from the chimney of a local chemical plant. As in earlier decades, the courts generally favored industry up to the late 1950s.[78] Indeed, the Ruhr Cooperative's emphasis remained the procurement of ample water for industry and not the purification of industrial or domestic wastewater. Even after water and public health officials in Düsseldorf demanded that the cooperative adopt more effective methods of wastewater treatment in 1952, the cooperative resisted. As a result, expenditures on wastewater treatment consistently lagged behind the demands placed on the Ruhr by an ever-expanding economy.[79]

Other factors also impeded any concerted plans for reducing water pollution in the Ruhr. The population remained heavily dependent on local mines and steel mills for employment, and not until sometime in the late 1950s did protests against environmental degradation in the region become sustained, a movement symbolized by the Social Democratic Party's "Blue Skies over the Ruhr" initiative. As in England, many government agencies could not afford to build the treatment facilities required by a booming economy and a host of new industries. Consequently, the Ruhr and Emscher further deteriorated, as did rivers elsewhere in the Federal Republic of Germany.[80] Referring to this period, wastewater expert Hendrik Seeger remarked, "Water pollution continued to get worse. The waste water treatment plants were not able to cope with the increasing population and the expanding economy, flourishing since the mid-1950s. Generally speaking, the peak of water pollution in West Germany was reached in the late 1960s."[81]

German Water Policy to 1990

As in England, the growth of modern environmental consciousness encouraged governments in the FRG to focus increasingly on water quality. Encouraged especially by Hans Dietrich Genscher, the FRG's minister of the interior, and by the adoption of environmental legislation in the United States, the government established a cabinet committee for environmental policy in July 1970 and shortly thereafter published a detailed program outlining future water policies in the FRG. Accordingly, the Bundestag passed a series of measures, the most important of which was an updated version of the Federal Water Law (known by its German acronym WHG). It required that water bodies "be protected as part of natural systems and as habitats for animals and plants." More to the point, for the first time, all entities wishing to discharge water into German rivers or streams had to obtain permission from state authorities to do so, and those permissions could be granted only if wastewater discharges met certain minimum standards, established at the *federal* level. Those standards covered a number of parameters, including measurements of dissolved oxygen, ammonia, nitrogen, heavy metals, and other substances.[82]

Despite this federal mandate, the actual structure of water management in the Ruhr did not substantially change. The Ruhr Cooperative remained in place, although it now assumed the responsibilities of the former Ruhr Dam Association, originally created in 1899. By the 1980s the Ruhr Cooperative remained organized as a public entity with members (and owners) representing more than sixty towns, seven hundred businesses, and thirty waterworks, while the cooperative itself owned and operated numerous wastewater treatment plants and reservoirs. The Emscher Cooperative remained in place as well, with responsibility for waste disposal along the badly contaminated River Emscher, a stream that was still a concrete-lined, open-air sewage canal. As stated above, the major change in legal status for both the Ruhr and Emscher Cooperatives was that they had to meet rigid federal water-quality and wastewater-discharge standards.[83]

In contrast to Great Britain, the Ruhr Cooperative has performed well since the early 1970s, especially in terms of reducing the worst forms of stream pollution. Between 1972 and 1992, its members voted to spend DM 2.1 billion on new sewage treatment plants and storm-water retention tanks. While the federal and EEC guidelines adopted in the 1970s have undoubtedly required these expenditures, the Ruhr Cooperative has been determined to provide enough clean water not only to serve a population of some 5 million people but also to encourage business expansion and development in the region. As a result, water quality has steadily improved since 1972 in terms of biochemical oxygen demand, heavy metals, and other pollutants.[84]

Much remains to be done. As in the industrial regions of northern England, the Ruhr economy has stumbled along over the past thirty years, with unprofitable mines and mills closing and higher-than-average unemployment levels.[85] This may partly account for the reduction in pollution as well. Yet in the 1990s there was some concern that the local populace would again be willing to accept high levels of pollution in exchange for new factories and new jobs. Since the 1990s, efforts to restore the Emscher have foundered on the immense investment such an endeavor would require.[86] Moreover, continually tightening EU and federal mandates will compel the Ruhr Cooperative to spend billions of euros in the next decade to alleviate pollution from nitrogen- and phosphorus-laden effluents. Nevertheless, the water management boards in the Ruhr have made remarkable progress compared to their predecessors in the previous century.

THE HISTORIES of Yorkshire and of the Ruhr offer important lessons about the role that government can play in solving environmental problems. In both regions, the Industrial Revolution brought radical changes to the landscape and massive environmental problems. Confronted with sudden shortages of drinking water and unprecedented quantities of wastewater, governmental institutions in Britain and Germany responded in critically different ways.

In England the political framework evolved slowly, with Parliament gradually incorporating more and more local entities into large RWAs. This system, as chaotic as it must have seemed to contemporary observers, fostered a rough but workable balance between consumers' need for plentiful clean water and a desire to limit the worst effects of stream pollution. Based originally on concerns about contagious diseases carried in human waste, the English system was transformed during the mid- and late twentieth century in order to address more insidious concerns about chemical pollutants, suspended solids, and declining oxygen levels. While far from perfect, the system worked because it encouraged major stakeholders to define and assert their rights under the law and under the supervision of regional agencies that embraced a variety of interest-group perspectives.

Until the 1970s, and notwithstanding the interwar years, England's political framework demonstrated the efficacy of a decentralized, pluralistic form of government. By that time, however, other factors had overwhelmed the ability of government to curtail water pollution. Even when Parliament reconfigured the RWAs to manage all aspects of the water cycle, the system proved unable to reconcile the demand for ample supplies of clean water with a reduction in river pollution. The increasingly rigorous water quality standards exacerbated this problem, as did the economic difficulties that plagued this sector throughout the late 1970s and 1980s. Absent the Thatcher government's making water quality a high priority and allocating funds to build expensive new treatment plants, the publicly owned system of water management was bound to fail. In a sense this verifies John Sheail's assertion that "a world of difference exists between having authority and using it to the greatest possible effect."[87] England's government had the authority but failed to use it to its full potential.

A comparison with Germany's Ruhr region confirms this observation. For the period up to about 1970, water policies in the Ruhr remained largely unchanged and geared almost exclusively to providing industry and households with ample clean water while preventing outbreaks of contagious disease. Stream pollution was an afterthought in this region. Moreover, the rigid cooperative system established by the state of Prussia in the early twentieth century—under crisis conditions—proved inferior to that of England's Yorkshire region for most of this period, at least in terms of protecting waterways from the worst kinds of pollution. Preventing pollution was never the intent of the politicians in Prussia who consigned control of the cooperatives to industry and other major polluters. The cooperatives served their intended purpose well, but local rivers and streams became badly polluted. Not until the 1970s, when the Ruhr Cooperative voted to fund a huge expansion in sewage treatment capacity, did water quality finally begin to improve.

The records of government agencies in Yorkshire and the Ruhr thus illustrate the relative effectiveness of different systems of water management under changing economic and environmental conditions. For the first two phases covered by this study, Yorkshire's water agencies performed admirably in comparison with those of Germany, while in the last phase, from 1971 to 1990, those in the Ruhr performed more effectively. While this judgment is, by definition, highly subjective, it does suggest a few final conclusions. Above all, the history of these two regions illustrates that modern environmental concerns—especially in the form of more rigid water quality standards—have also encouraged a more uniform approach to water management. In both Great Britain and Germany, for instance, water management has become more centralized since 1970. Under a series of British governments, Parliament gradually consolidated the old water boards into a series of large water authorities, before selling them off in the late 1980s. Even then, these private agencies remained subject to a regimen of national and EU water quality standards.[88] Likewise, in Germany, legislation passed during the 1970s imposed nationwide water quality standards, and Germany was also subject to EEC directives. Ultimately, however, even this framework of legislation, standards, directives, and national ministries has proved insufficient to guarantee both economic growth and clean waterways.

8

Saving the Rhine
Water, Ecology, and Heimat *in Post–World War II Germany*

THOMAS LEKAN

Celebrated by the nineteenth-century Romantics for its natural beauty and nationalist symbolism, Germany's Rhine River was widely known in the post–World War II period as the "sewer of Europe."[1] A toxic brew of improperly treated organic waste from municipal treatment plants, nitrates and pesticides from agricultural runoff, phenol and chloride salts from coal mining, heavy metals from chemical manufacturing, and thermal pollution from nuclear power plants had made its water unsafe to drink or swim in and had destroyed its once-thriving fisheries. While the Rhine contains only 2 percent of the Western world's volume of water, its banks became the site of nearly one-fifth of the world's chemical plants, effluents from which posed a constant risk for the 20 million people dependent on the river for their water supply.[2] By the late 1950s it was clear that the Rhine could no longer fulfill its dual role as a source of potable water and a dumping ground for municipal and industrial wastes.

The river's future appeared much brighter by 1996, however, when the French newspaper *Le Monde* declared the Rhine to be the "cleanest river in Europe," a judgment confirmed by several international and national monitoring agencies.[3] National standards for regulating pollution established by the German government in the 1960s and 1970s, as well as the multilateral Rhine Action Plan of 1987, saved the river from ecological extinction. The action plan was signed a year after an environmental disaster occurred near Basel, Switzerland, in which thirty tons of pesticides were washed into the

Rhine during the effort to extinguish a fire at the Sandoz chemical factory. So when German newspapers reported in December 1990 that a hatchery-raised salmon had been caught in the Sieg River, a tributary of the Rhine, it seemed that Europeans had learned quickly from the Sandoz tragedy and were partly on their way to restoring the Rhine's ecological integrity.[4]

The precipitous decline and partial recovery of the Rhine in the postwar era mirrored broader changes in German environmental policies and attitudes over the same period. Historians and journalists have usually identified the 1970s and 1980s as the decades in which modern, scientific awareness of the Rhine's ecological woes and other environmental issues replaced supposedly regressive, agrarian-romantic forms of nature conservation.[5] Massive protests against the construction of nuclear power plants at Wyhl and Breisach in the 1970s and public outcry over the Sandoz incident galvanized popular opinion about the intolerable condition of the Rhine's waters, spurred the formation of alternative political lists and local Green Party affiliates, and underscored the need for state regulation of toxic emissions in the interests of public health.[6]

This focus on the 1970s and 1980s as an environmental watershed has been challenged in recent years. Some historians argue that West Germany's heyday of environmental activism was in many respects a delayed reaction to scientific and journalistic warnings about ecological degradation dating back to the 1950s and 1960s. Whereas some scholars have depicted this period as one in which society suffered from a "fifties syndrome" that valorized consumerism and fostered apoliticism, scholars such as Raymond Dominick, Sandra Chaney, and Monika Bergmeier have shown that the 1950s and 1960s were instead a "gestation period" for an increasingly sophisticated, scientifically informed awareness of environmental problems. In some localities the period witnessed renewed popular protest against infrastructure projects that threatened the natural environment.[7]

As I show here, this alternative periodization for the rise of environmental awareness in the Federal Republic offers new insights into postwar efforts to save the Rhine. Long before the antinuclear protests at Wyhl mobilized public attention, states along the Rhine had formed organizations dedicated to investigating the sources and effects of the river's pollution. In addition, natural scientists had predicted the river's impending ecological collapse, and newspapers and magazines had routinely published detailed accounts of the *Wassernot,* or water emergency, in Western Europe. Environmentalists of the 1970s and 1980s did help to push through new water management legislation, and they pioneered use of the term *Umwelt,* or environment, to describe the manifold relationships between human beings and their physical and organic surroundings. However, these legal measures and discourses were not a spontaneous reaction to accumulated environmental stresses that came to light only in 1970 but the implementation of proposals first advanced between 1955 and 1965, many of which addressed pollution control along the Rhine.

Pushing the origins of ecological attitudes toward the Rhine and other waterways back to the 1950s, however, raises crucial historiographical problems about the development of postwar West German environmentalism. How did the largely aesthetic and nationalistic forms of nature conservation (*Naturschutz*), homeland protection (*Heimatschutz*), and landscape preservation (*Landschaftspflege*) that predominated in Germany before 1945 evolve into the scientifically grounded, systems-oriented concept of environmentalism developed in the postwar period? How did German conservationists and ecologists overcome the tainted association of Romantic naturalism with Nazism and find an effective strategy for promoting environmental ideals in the wake of the Third Reich? The Rhine offers a highly informative case study of the continuities and discontinuities in German environmental politics, as the fate of the river and its surrounding landscape in the industrial era aroused intense debate in both the prewar *Naturschutz* movement and the postwar environmental imagination. I argue here that nature conservationists' growing interest in the Rhine's toxic contamination (or *Verunreinigung*) and increasing reliance on scientific data and ecological models to understand the systemic impact of such pollutants constituted a significant conceptual departure from the environmental concerns of the Weimar and Nazi eras. Before 1945, environmental concern about the Rhine tended to focus on the aesthetic disfigurement (*Verunstaltung*) of the river's surrounding landscape rather than on contamination of the river itself.[8] Conservationists tried to set aside what was left of "pristine" nature or visually captivating landscapes for recreation, cultural refinement, or scientific study, but they ignored the unseen, insidious effects of industrially produced toxins. After 1945, conservationists continued to advocate for the creation of nature parks and preserves, but they also condemned industrial polluters and sought out legal and institutional means to curb the discharge of chemical effluents into the water, air, and soil.

While the theoretical foundations of West German nature conservation began to shift as soon as the war ended, the moral critique, social elitism, and cultural meaning of environmentalism did not change so readily, revealing a disjunction that exposes important discursive and institutional continuities across the 1945 divide. Though state officials, hydrologists, ecologists, and environmental groups strenuously avoided the xenophobic references to an ethnically homogeneous people (*Volk*) and living space (*Lebensraum*) that had characterized Nazi environmental discourse, the moral geography of homeland (*Heimat*) and the tradition of conservative, educated, middle-class cultural criticism (*Kulturkritik*) that had framed conservation practice in the late nineteenth and early twentieth centuries continued into the postwar era. Just as many conservative elites within West Germany blamed the atomized "mass man" for the triumph of Nazism, so, too, did nature conservation organizations, state officials, engineers, and journalists readily attribute ecological problems to a spiritual malaise within West Germany and Europe as a whole, a hollow "materialism" that fueled public apathy toward the natural *Heimat*.[9]

This tendency to attribute environmental decline to moral weaknesses in West German society, rather than to the exploitive character of industrial capitalism or state-sponsored brute force technologies, underscores the critical role that cultural discourses played in framing the cumulative evidence of the Rhine's ecological decline.[10] As recent studies from the fields of political ecology, environmental justice, and environmental anthropology have shown, the perception of threat from pollution rarely correlates directly with quantitative levels of contamination or even direct sensory experience. As Peter Thorsheim has emphasized in connection with the London fog disaster of 1952, "one must do more than identify the environmental conditions and toxic chemicals that caused illness and death. Equally important are the attitudes, ideologies, and perceptions that led to the creation of these pollutants and that structured people's understanding of their effects."[11] Thorsheim's work is part of a new generation of scholarship that analyzes not just the scientific and economic effects of pollution but its cultural meanings as well. This scholarship has verified empirically the theoretical insights of anthropologist Mary Douglas, who in her landmark work *Purity and Danger* showed how societies' ideas about dirt, waste disposal, and hygiene have often covertly sanctioned particular beliefs about the moral universe and reinforced claims to cultural authority and political power. In Douglas's view, fears of contamination often reflect anxiety about changing social hierarchies and cultural norms. The elimination of wastes, in turn, becomes a creative rather than negative act, a "positive effort to organize the environment ... [,] to relate form to function, to make unity of experience," and to "atone" for past sins.[12]

In the case of the Rhine, water contamination magnified fears among bourgeois nature conservationists, sanitary engineers, and state officials about the cultural decline and "Americanization" of European society, in sharp contrast to the mood of optimism that pervaded other sectors of West German society during the "miracle years" of the 1950s and 1960s.[13] For these groups, postwar attempts to restore the Rhine's ecological integrity and to purge the "excrements of industrial civilization" dovetailed with the ethical rejuvenation and cultural reinvention of Germany in the wake of war and defeat. Members of these groups imagined water purification as a stepping stone to social reconstruction, thus adding a moral urgency to West Germany's environmental debate well into the 1980s. This element was missing from responses by other riparian states to the environmental crisis of these decades. Efforts to rehabilitate the Rhine also bolstered the symbolic reinvention of *Heimat* as a locus of ecological concern and regional identity while offering a sharp contrast to Nazi centralization. International observers portrayed both the war and water pollution as symptoms of the nihilism and materialism that still threatened to engulf European, not just German, civilization. Indeed, the river's historical role as a contested frontier between France and Germany made cooperation to resolve its ecological problems a litmus test for European efforts to integrate and come to terms with the recent, violent past.

Wassernot! Water Pollution and Ecological Anxiety in the Postwar Era

Postwar awareness about the extent of the Rhine's ecological degradation began in April 1946, when the Netherlands registered a formal complaint with the Central Commission for Rhine Shipping. The complaint requested international arbitration over the condition of the Lower Rhine, which the Dutch claimed was vastly more polluted than it had been before the war. Dutch delegates noted with alarm that the number of fish caught in the river had dropped sharply since 1945. They feared that fishery declines might signal a broader contamination of the river, on which the Dutch depended for 61 percent of their water for household consumption and agricultural needs. Allied commanders and German officials attributed this decline to the destruction of sewage treatment plants by Allied bombing and confirmed that in many places raw sewage and industrial wastes, including those from the notoriously polluted Emscher River, flowed untreated into the Rhine.[14]

Seeking a more permanent solution to water pollution regulation, the Dutch government eventually called on the three other nations along the Rhine—France, Switzerland, and the newly created Federal Republic of Germany—to form an advisory council charged with improving the Rhine's waters. This action led to the establishment of the International Commission for the Protection of the Rhine against Pollution (hereafter the Rhine Commission), which met for the first time in 1950 but was formally chartered in 1963. The original group of four was joined by Luxembourg, and these five states committed themselves to monitoring and reporting on the Rhine's environmental status, holding regular consultations, and developing and supporting policies that would stem further decline of the river. The states agreed to measure pH levels as well as the amounts of free oxygen, ammonia, nitrates, chlorides, phenol, and radioactivity in the Rhine's waters and to publish the results in a series of annual reports during the 1950s.[15]

The monitoring reports revealed a staggering degree of environmental degradation, far worse than anything that could be explained by a wartime lapse in wastewater treatment. After only three annual reviews the commission judged the Lower Rhine to be so laden with pollutants that it would require emergency pollution prevention measures to ensure the safety of Dutch and German water supplies. Separate investigations conducted in North Rhine–Westphalia in 1959 revealed an enormous increase in toxins as the river flowed through this state's heavily urbanized and industrialized corridors. These toxins, which entered the Rhine at Duisburg, included effluents from the chemical and pharmaceutical plants near Cologne and from the coal mines and foundries of the Ruhr region. As the Rhine wound its way between the town of Honnef and the Dutch border, potassium manganate levels increased 7 percent; chloride salts, 72 percent; phenol, 250 percent; iron, 300 percent; and germs such as fecal bacteria, an unprecedented 493 percent.[16]

State and public concern in Germany about the condition of the Rhine's waters was certainly not new. Even during the heyday of Rhine romanticism in 1834, Samuel Coleridge commented on the foul smells emanating from the river with the quip, "The River Rhine, it is well known, / Doth wash your city of Cologne, / But tell me Nymphs, what power divine, / Shall henceforth wash the river Rhine?"[17] As urbanizing municipalities along the Rhine exhausted groundwater supplies and accumulated more human wastes than septic tanks could handle, they relied more heavily on the river for both drinking water and waste disposal, increasing the risk of cholera and other waterborne diseases.[18] In 1897 the town of Mannheim's decision to discharge untreated sewage into the Rhine only eight miles upstream from Worms prompted lawsuits and an acrimonious exchange in regional newspapers. Worms was eventually forced to revert to groundwater wells for its municipal water needs after state and local authorities refused to compel Mannheim to install an effective wastewater treatment facility and instead advised Worms to construct a new filtration system. Mannheim's victory led many other cities to ignore calls for effective wastewater treatment; as late as 1956, water authorities in North Rhine–Westphalia noted that at least twenty-nine cities along the Rhine in their state lacked adequate sewage treatment.[19]

Industrial pollution posed a more serious threat to the Rhine. The growing number of chemical plants that used the river for production, heating and cooling, wastewater disposal, and transporting finished products increased its toxic load, since the manufacture of dyes, explosives, fertilizers, acids, alkalis, and paper released phosphorus, nitrogen, heavy metals, and, later, chlorinated hydrocarbons into the river.[20] Yet well into the twentieth century most public health experts continued to focus on municipal organic wastes, rather than industrial effluents, as the most dangerous contaminants in the Rhine. Water experts lacked the equipment and expertise to effectively test for contaminants, and the long-term effects of chemical toxins on disease and mortality rates among humans remained largely unknown. Scientists and public health authorities before World War II did express concern about the wastes dumped into the Lower Rhine by Germany's military-industrial complex in the Ruhr coal mining region. A government study in 1913, for example, showed that sewage and residue from mines and metal works had contaminated the Ruhr's riverbed and impaired the natural biodegradation process.[21]

The condition of the Emscher River, which flowed through the coalfields of Westphalia before emptying into the Rhine, was even worse. Inadequate water laws led the Prussian government to form the Emscher Cooperative in 1904, a "riparian cooperative" dominated by industrial water users that transformed the Emscher into a shortened, concrete-lined effluent canal.[22] Waste from 100 factories, 150 coal mines, and 1.5 million people flowed into the Emscher, creating a reeking, steaming, oozing black mass of raw sewage, chloride salts, phenol, cresol, cyanide, zinc, mercury, and chromium. This toxic sludge wound its way toward the Rhine, prompting one Reichstag member in the 1920s to call the Rhine "Der Höllenfluss"—the River of Hell. The Emscher Cooperative did help to install wastewater and sewage treatment plants in the region, including more than a hundred of Karl Imhoff's so-called Emscher wells to remove solids from wastewater. Biological treatment processes were also available but were considered too costly for widespread use until after 1918. Neither Imhoff's technology nor biological treatment, however, could deal effectively with chemical pollution. Of the chemicals in the Emscher and Rhine waters, foul-smelling phenol was perhaps the most recognizable, since it permeated the fatty tissue of Rhine salmon and mayfish, thus making these species inedible. Dephenolization techniques pioneered in the 1920s promised to prevent this noxious chemical from contaminating fish, but high costs meant that by 1956 the Rhine region contained only twenty dephenolization plants, barely enough to purify one-half of the phenol-contaminated wastewater from the Emscher.[23]

The stepwise canalization of the Rhine in the nineteenth and twentieth centuries exacerbated such pollution problems and raised the risk of flooding by converting the alluvial plains that had previously acted as a sponge for high waters and a filter for toxins into cultivated farmlands and residential areas. Canalization also tended to lower the groundwater table and desiccate the surrounding landscape, which made residents still more dependent on the increasingly contaminated Rhine.[24] Biologist Reinhard Demoll noted in 1952 that a stretch of the Rhine riverbed near Breisach had sunk almost seven meters since the early nineteenth century, which meant that it had deepened at three hundred times the normal rate. According to Demoll, more than three thousand hectares on the edge of the Black Forest resembled a desert due to the sinking water table, costing farmers almost DM 100 million over a fifteen-year period.[25]

Despite the seriousness of these pollution and hydrological problems, late nineteenth-century water experts such as Max von Pettenkofer believed that rivers had virtually unlimited "self-cleaning" capacities, while others argued well into the twentieth century that certain stretches of rivers (the so-called *Opferstrecke*) could be sacrificed to industry without damaging the rest of the river.[26] Nature conservationists' depictions of the river during the 1920s and 1930s paralleled this generally optimistic assessment of the harmonious balance between nature and industry along the Rhine. Indeed, before World War II nature conservationists and *Heimatschutz* advocates tended to ignore signs of the Rhine's ecological deterioration in favor of a preservationist ethic that focused on the aesthetic beauty and nationalist or regional significance of the river's cultural landscape.

In the 1950s and 1960s, as scientific analyses and journalistic reports drew attention to the flood of new pollutants overwhelming what was left of the river's biodegradation capacities, the ecological integrity of the river's waters became a matter of public concern. A growing number of West Germans and other Europeans along the Rhine's banks began to fear that the entire river—Lower, Middle, Upper, and High Rhine—would soon become a sacrificed stretch.[27] In the early 1950s, for example, those who fished in the Rhine's waters complained of a plague of oil slicks (*Ölpest*), the noisome by-product of Germany's rapid motorization in that decade and the conversion of thousands of households from coal to heating oil. Newspapers and environmental organizations reported that accidents involving petroleum products could have lethal results; a single liter of gasoline, they noted, could make a million gallons of water unpotable.[28] A sharp decline in the number and diversity of invertebrate megafauna in the river registered the cumulative impact of these contaminants; of the eighty indigenous species of crabs, mussels, and snails that inhabited the Rhine in 1915, only forty-two species remained in 1956. The organisms that took their place were often exotic species resistant to salt and pollution, posing an additional threat to those species that remained viable in the Rhine watershed.[29] By most accounts, the Rhine stood on the brink of ecological collapse by the late 1960s, its regenerative capacities overwhelmed by industrial filth.[30]

By far the most sinister threat to the Rhine and other watersheds, at least in the eyes of many West German government officials, journalists, and citizens, was radioactive fallout. Germans knew that if the Cold War turned hot, their country would be the major battleground in a campaign of nuclear annihilation, a prospect that made them particularly susceptible to fears of radioactive contamination.[31] American and Soviet weapons testing in the 1950s produced dangerous concentrations of radionuclides in rain and surface waters over Central Europe, resulting in levels of radiation in drinking water that were up to one hundred times above normal in some areas.[32] Though both government and popular opinion generally favored the use of atomic energy, stories about radioactive contamination reinforced fears about other environmental dangers and drove home the message of interdependency, which was part of the emerging "ecosystem model" of nature. As Raymond Dominick notes, "While reading about the radioactive rain that fell on Germany, the public learned that environmental threats might be distant in origin, invisible in operation, delayed in damage, and extremely long-lived in their menace..., causing both cancer and birth defects."[33] These atomic-age anxieties drove West Germany's cadre of "apocalyptic" environmental writers in the 1950s and 1960s to issue warnings of an impending ecological collapse from nuclear and chemical contamination, overpopulation, and strains on global natural resources. Reinhard Demoll (founder of the Kampfbund gegen Atomschäden, or Fighting League against Atomic Injuries), Günther Schwab (founder of the Weltbund zum Schutz des Lebens, or World League for the Defense of Life), and Bodo Manstein pointed to atomic fallout and radioactive wastes as some of the most insidious problems facing postwar societies. They ridiculed the "peaceful" use of the atom for inexpensive and "clean" power generation.[34] Their analyses helped to shift the locus of Germans' environmental concern away from a predominantly visual and aesthetic view of landscape preservation to an ecologically informed and holistic sensibility, commonly referred to in the 1970s as the *Umwelt,* or environment, the protection of which was known as *Umweltschutz.*

Thermal pollution from nuclear reactors raised additional fears about the environmental costs of civilian atomic energy. Scientific studies reported that wastewater from dozens of reactors along the Rhine and other watersheds in West Germany that used river water for cooling could collectively raise riparian water temperatures to forty or even fifty degrees centigrade, high enough to kill all organisms in the water and to produce vapors so dense they would blot out the sunlight.[35] The construction of cooling towers at several facilities assuaged the concerns of many conservationists and fishing enthusiasts about nuclear facilities, but by then some groups on both sides of the river had had enough and began organizing some of the most spectacular protest actions of the 1970s. With the slogan "Better active today than radioactive tomorrow," French, German, Swiss, and Dutch environmentalists joined together to resist the nuclearization of the Middle and Upper Rhine. The establishment of the Rhine Valley Action Committee successfully blocked or delayed construction of reactors at Marckolsheim, Breisach, Wyhl, and Kalkar. Their success emboldened other environmental and green groups to protest nuclear energy and comparable environmental abuses in the 1980s and 1990s.[36]

In the environmental discourse of the postwar decades, water thus served as the most important physical and symbolic medium for transporting, detecting, and magnifying fears of radioactive and other toxic contamination. As the substance indispensable to life on earth, the waters of the Rhine and other rivers and streams enabled Germans to recognize the systemic connections between humans and the biosphere as well as their vulnerability to industrial filth. While water sustained life, it also conveyed invisible and potentially deadly toxins across political boundaries and geographical barriers, whether falling from the skies or flowing into wells and reservoirs. The complexity of water pollution also challenged Germans' long-standing belief in technology as an effective means of controlling the environment. As an information newsletter for the German Association of Gas and Water Experts explained in October 1959, water researchers' attempts to understand the impact of toxins on the Rhine's dynamic riparian system were complicated by "the many-sided mutual dependencies in the organic world that shape the river's household. . . . The dangerous consequences are still too little understood because of the mutual influence of the various substances in the human body, e.g., the long-term effect of small dosages."[37] Other water experts lambasted the notion of "tolerable exposures to the various poisons that were being pumped into the streams and rivers. Measurements were notoriously unreliable ... and the poisons worked in cumulative and synergistic ways."[38] Small wonder, then, that fears of an impending *Wassernot* (water crisis), fueled by a combination of overconsumption and pollution and centered on the Rhine and its tributaries, became a dominant environmental concern of the postwar period.

Ecology did not emerge as a legitimate biological subfield in German universities until the late 1970s. Still, in the 1950s an "ecologistic" understanding of the disposal, transmission, bio-accumulation, and health impact of waterborne toxins did contribute to a comprehensive, holistic framework for understanding why water pollution threatened not only wilderness and rural areas but also urban and suburban landscapes. Scientists, politicians, public health experts, and nature conservationists recognized that protecting greenfields, woodlands, or heaths as unsullied spaces apart from human communities would no longer suffice; what was needed were conceptual tools for managing the physical and biological foundations of the human-shaped environment. In this context, field biologists and other natural scientists attuned to ecological questions began to enter the public sphere and to sound the alarm about the implications of ecosystem research for human communities.

The limnologist August Thienemann, for example, pioneered the holistic study of riparian systems in the interwar period, long before ecology was a recognized branch of biological study.[39] As early as 1911, Thienemann had investigated the effect of drought and pollution on fish populations in the Ruhr, detailing the many effects of industrial toxins on aquatic species.[40] In his treatise *Man as Member and Shaper of Nature* (*Der Mensch als Glied und Gestalter der Natur*), published in 1944, Thienemann extended Karl Möbius's concept of biocoenosis (from 1877) to argue that plants and animals in a particular range of soil, climate, and water conditions formed an organism of a higher order. The resulting organism, or *Lebensgemeinschaft*, interacted with and depended on a particular configuration of climatological and topographical factors, the *Lebensraum*. Thienemann analyzed lakes and rivers as intricate webs of plants, animals, and energy flows that, in the absence of disturbance, tended toward dynamic but stable balance.[41]

Thienemann's use of the geographer Friedrich Ratzel's term *Lebensraum* to describe environmental conditions and *Lebensgemeinschaft* to describe higher-order organisms had uncomfortable associations with National Socialist geopolitical rhetoric. After 1945, therefore, Thienemann abandoned these terms and began to use the word *Umwelt* to describe the manifold environmental conditions that shaped the lives of organismic communities, decades before it became a household word in Germany.[42] Thienemann's ecological perspective also achieved new recognition in the postwar era. In a *Deutsche Zeitung* article in 1956, Thienemann argued that every waterway in Germany should be considered "a unitary whole, a branch of Nature, a 'microcosm,' a world in miniature, bound to the surrounding landscape by inherent mutual influences."[43] Industrialization and urbanization had disrupted the natural homeostasis of such a system; as he warned in 1959 in his memoir, modern technology had led "the use of waterways to become misuse!"[44]

Thienemann's transformation from a relatively obscure interwar ecologist into a prominent scientist after 1945 paralleled the growing influence of scientists and physicians such as Demoll and Manstein in the West German public sphere. It also signaled the professionalization and, by the 1960s, the technocratization of environmental issues in the postwar era. This tendency accelerated as new conservation organizations emerged, such as the Alliance for the Protection of German Waters (Vereinigung deutscher Gewässerschutz, or VDG), whose publications touted ecological and technical solutions to Germany's water management crisis. Founded in 1951 by those who engaged in commercial and recreational fishing in Frankfurt to arouse public interest in water resource issues, the VDG lobbied hard for a national water protection law that would allow for large-scale management of water resources. The VDG also served as a meeting point for constituencies with diverse interests in water usage—scientists, civil servants, elected officials, nature conservationists, industrial representatives, tourism promoters, and farmers—and by 1954 it numbered almost 1 million members.[45]

Speakers at the VDG's symposia and exhibitions included leading politicians and scientists. At the alliance's conference in 1959 the federal minister for atomic energy and water supply, Siegfried Balke, noted that Germany's environmental crisis had become "life-threatening" due to the greed and corruption rampant in the nation's postwar economy, which he referred to as "excrements of civilization."[46] In order to educate the public, the alliance published notable works on Germany's water problems, including several by Erich Hornsmann, one of the country's leading hydrologists. Hornsmann's *If We Did Not Have Water* (*Hätten wir das Wasser nicht*), published in 1957, asserted that Germany was already in the midst of a water emergency that threatened to stall postwar economic recovery and shake the foundations of German society. With more frequent bathing and industrial needs, he noted, the demand for water had increased during the 1950s while the usable supply of water had decreased, so that water had become a *Mangelware*, a commodity in short supply.[47] By the early 1970s the alliance had sponsored another thirty-one titles in its series and become a leading force in drafting a water management law enacted in 1957 and subsequent legislation dealing with detergents and oil pollution.[48]

Like scientists and environmental organizations, journalists were in the vanguard of public awareness about the Rhine in the 1950s and 1960s. They used the newly reestablished freedom of the press to draw citizens' attention to environmental problems, to pressure politicians to pass pollution control legislation, and even to sell newspapers and magazines through sensationalized headlines and images of impending environmental disasters. "The Rhine Still Stinks!" noted a *Bildzeitung* article published in 1960 and calling for the Bundesrat to pass a more stringent national water law, and in 1970 *Die Welt* reported that "20 Million Drink Water from Germany's Largest Sewer."[49] In November 1959, after West Germany had faced its driest summer since 1890 and implemented widespread water rationing, *Der Spiegel* printed an eleven-page feature on the country's *Wassernot*. The article focused on water scarcity and pollution problems in the Federal Republic, graphically depicting the causes of water overconsumption, the sources of water pollution, and the hydrological consequences of streambed regulation.[50]

The article charged municipalities and the chemical industry with despoiling the country's already dwindling water supplies, noting that industries returned 57 percent and households 83 percent of their wastewater to the country's surface waters in a polluted condition. This process posed a dire threat to cities such as Düsseldorf, which still depended on the Rhine for 80 to 85 percent of its water supply.[51] The *Der Spiegel* report pointed out that West Germany's confusing patchwork of water regulations and its ineffectual national water law of 1957 were partly to blame for the country's water crisis. Subsequent letters to the editor castigated the government for spending limited funds on "atomic armaments" instead of civilian needs.[52] Local Rhineland journalists were also attuned to the new climate of ecological anxiety. The authors of an article printed in *Düsseldorfer Nachrichten* in 1952 wrote, "The mighty river is ... sick ... no longer in a state where it can digest the dirt and masses of feces fed into it.... The poisonous substances phenol, zinc, chromium that are allowed to enter the river as wastewater in large quantities needed to be removed mechanically beforehand. The weakened self-cleaning powers of the river, which has been transformed into a cesspool, are no longer able to master them."[53]

The West German public proved a receptive audience for these grave warnings of impending environmental disaster. Statistics about the extent of Rhine pollution and predictions of a water emergency appeared regularly in newspapers, films, and magazine articles in the 1950s, creating a popular ecological understanding that incited citizens' initiatives and environmental protest two decades later. The grotesque fish kills of the postwar decades confirmed hydrologists' and scientists' declarations that Germany's rivers were at the limit of their carrying capacity. A study commissioned by the Alliance for the Protection of German Waters showed that between 1949 and 1952 more than one hundred human-induced fish kills had occurred in the Federal Republic each year.[54] In one incident in 1957, the swollen cadavers of salmon, whitefish, and barbel washed up along a twenty-kilometer stretch of the Ahr River, a scenic tributary of the Rhine popular with sport anglers from nearby Bonn. The cause: a highly concentrated sodium-cyanide agent, released from a coal purification plant and made deadly through interaction with trace amounts of naturally occurring ammonia in the river water.[55] The effect was catastrophic, not only for the poisoned fish but also for the entire array of bacterial microorganisms in the water and sediment. In 1969 millions of fish died as a result of the deadly pesticide Thiodan in Rhine waters. It was an environmental scandal that caused even East Germany's Socialist Unity Party to declare the Rhine a "Fluss des Schreckens"—a river of horrors.[56]

The fish kills demonstrated the concept of ecological interdependence and the slow death of Mother Nature with a vividness and immediacy that scientific studies and news reports could never match. In the case of the Ahr, invisible pollutants producing synergistic effects with ammonia in the water showed that piecemeal regulation of effluents from individual factories could not safeguard the public against waterborne disasters. Sodium cyanide caused deaths all along the food chain, from the microbes in the sediment to the fish predators at the top. This situation prompted growing concern about the human link in this chain; the campaign by Rhine fishing interests to save the river by suing polluting industries became not merely a "minor battle for their own livelihood" but a "fight for the preservation of the people's health" as well.[57] Their efforts were sadly in vain. Of the forty-two species of fish found in the Rhine's waters in 1880, only twenty-three remained in 1975, and the salmon catch had plummeted from around 225,000 in 1885 to nearly zero in 1945.[58] Commercially valuable salmon, mayfish, and sturgeon were thus among the many environmental victims of West Germany's economic miracle.

Beyond *Naturschutz*: The Moral Aspects of Water Quality in the 1950s and 1960s

By the late 1950s, therefore, the foundations of a modern, ecosystemic understanding of the Rhine's water problems were in place. Those foundations were reinforced by the presence of persistent and toxic industrial pollutants, scientific evidence of an impending *Wassernot,* and media attention to massive fish kills, radioactive contamination, and tainted water supplies. Such problems challenged Germany's existing *Naturschutz* organizations—which had traditionally focused on "natural monument" protection, cultural landscape preservation, or regional planning—to recast their aims in the postwar era. As Raymond Dominick has shown, traditional nature conservationists were not the vanguard of an ecologically based, globally oriented environmentalism in the 1950s and 1960s. Favoring cooperation rather than confrontation with government and industry, nature conservationists shunned the antinuclear tactics of Demoll's Fighting League and Schwab's World League for the Defense of Life, which preferred press conferences, referenda, citizens' initiatives, and lawsuits to behind-the-scenes persuasion of political leaders and close cooperation with traditional nature conservation groups and state agencies.[59] *Naturschutz* leaders had also often overlooked urban environmental concerns, such as pollution and overpopulation, and concentrated instead on preserving the visual harmony of rural landscapes or protecting characteristic features of the natural *Heimat,* such as heaths, hardwood forests, or rock formations. Some environmental leaders, such as Bavarian state nature conservation commissioner Otto Kraus, even favored the development of nuclear energy as a means of preventing further construction of hydroelectric dams, which despoiled scenic waterways.[60]

By the 1950s, however, nature conservationists could no longer ignore urban pollution. Though many conservationists continued to campaign for the establishment of nature preserves and scenic landscapes for recreation and study, others viewed water pollution prevention as a way to broaden their environmental engagement, attract new supporters, shed their reputation as nostalgic aesthetes, and assume a new role as modern environmental managers. In the 1950s and 1960s, conservationists often incorporated pollution control in their concept of regional planning, or *Landespflege,* which envisioned comprehensive management of human habitats.[61] As Sandra Chaney explains, even though reporters and newspapers of the 1970s claimed that "Grandpa's *Naturschutz* is dead," conservationists' efforts in the 1950s and 1960s did expand West Germans' environmental sensibilities and ensure that

Naturschutz would remain an important branch of environmental protection (*Umweltschutz*) in the Federal Republic.[62]

This shift in priorities undoubtedly reflected the unprecedented scale and scope of the ecological crises along the Rhine and elsewhere, as well as growing awareness of international trends in conservation and preservation. The technological optimism and materialism that accompanied the "economic miracle" and the triumph of a mass consumer society challenged West German conservationists, as it did their counterparts in other Western industrialized countries, to develop innovative justifications for nature protection in an increasingly democratic, urbanized, and affluent society. In such a society visions of both political freedom and the good life depended on constantly rising prosperity. In the United States, a similar expansion in environmental awareness beyond wilderness protection and natural resource management occurred after the publication in 1962 of Rachel Carson's *Silent Spring*, which alerted Americans to the growing threat to human health and wildlife populations posed by DDT and other pesticides. *Silent Spring* had appeared in a German translation by 1963, prompting West Germany to ban DDT in 1972. West German environmental agendas did not merely copy American trends, however. The country produced its own cadre of environmental jeremiads in this era, and the transition from conservation to the environment articulated larger sociocultural anxieties unique to the German experience of "coming to terms with the past."

In line with this holistic perspective, conservationists' environmental and cultural representations of the Rhine in the 1950s and 1960s merged landscape preservation discourses of the prewar era with a scientific and managerial focus on enhancing the river's water quality and ecological viability. Before 1945 conservationists had viewed environmental degradation through a nationalist or regionalist lens, arguing that the "disfigurement" of the countryside would surely erode Germany's distinctive national character, alienate urban workers from their homeland, and accelerate moral decline. Embracing the work of Wilhelm Heinrich Riehl, a student of Ernst Moritz Arndt and conservative social theorist of the mid-nineteenth century, conservationists saw each cultural landscape—whether local, regional, or national—as a reflection of ethnic identity and a repository of collective memory that anchored the community in space and time.[63] Following Riehl, many conservationists portrayed the enjoyment of meadows and rock formations as a form of patriotic devotion and a palliative for the grime and stress of city life. The Rhine had played a crucial part in this nationalization of landscape; when visiting sites where Napoleon's armies had been driven back into

France and when touring the castle ruins, vineyards, mountains, and half-timbered cottages in surrounding areas, Germans were invited to visualize their country as an ancient, primordial entity created by a unique cultural landscape, rather than a political experiment born of mid-nineteenth-century power politics.[64] Before 1945, in other words, conservationists worried more about the effect of industrial modernity on the character of *Heimat* than on the country's ecological stability.[65]

During the late 1920s this discourse of "land and people" (*Land und Leute*) became increasingly modernist and technocratic. Leading conservationists worked with regional planners to integrate factories and apartment complexes into the landscape and to facilitate industrial laborers' access to green spaces and recreational facilities in order to enhance industrial productivity.[66] The National Socialists adopted this planning ethos and made it a cornerstone of the Reich Nature Protection Law (*Reichsnaturschutzgesetz*, or RNG) of 1935, which on paper was one of the world's most stringent nature conservation laws. The law reflected the initial influence of *Heimatschutz* leaders on National Socialist cultural policy. Conservationists placed great hope in the RNG, expecting the Nazis to sweep away profit-driven "liberal materialism" and to revive the eternal values of nature and *Heimat*; they soon found, however, that in the Third Reich, military exigency trumped nature conservation.[67]

In the period between 1937 and 1941, for example, Rhenish conservationists and local officials developed plans for designating the Middle Rhine Gorge as a protected landscape under section 19 of the RNG in an attempt to reverse the "disfigurement" of this scenic treasure. The plan restricted commercial or industrial development to a few designated areas and called for the removal of unsightly billboards and garish building colors. It also directed local officials to consult nature protection authorities before allowing any major development, including road or railway construction, and specified that only "native" ("*bodenständig*") building materials and vegetation be used for new construction projects and gardens.[68] Planners recognized oil slicks and rancid odors from phenol emissions as forms of "disfigurement" but offered no measures for regulating pollution. Like earlier forms of *Heimatschutz* and *Landschaftspflege*, the Rhine landscape protection plan emphasized visual harmony over ecological restoration.[69] In line with Nazi ideals, it also bolstered racialized views of the Rhine's cultural landscape. When the Rhine province's Nazi governor referred to the river in 1941 as the "pulsating life's vessel" of the nation, he depicted the Rhine and its picturesque landscape as a reflection of essentialist ethnic traits and characterized nature appreciation as a pathway to racial consciousness.[70] Conservationists continued to develop plans for a Middle Rhine Gorge protection zone into 1941, but war on the eastern front soon placed the "Battle for Production" far ahead of nature conservation on the Nazi regime's list of priorities.

Given this problematic association between landscape preservation and Nazism, it is hardly surprising that conservationists sought to move beyond *völkisch* justifications for environmental protection after 1945, even though they did not renounce the environmental achievements they had attained as a result of the conservation law of 1935.[71] Instead, they signaled their break with prewar conservationism by incorporating emerging ecological perspectives, water pollution concerns, and dire warnings about the threats to human —not just German—habitat into their vision of environmental reform. The film *Nature in Danger* (1952), which was commissioned by the Bavarian conservation officer Otto Kraus, identified a variety of environmental abuses, from stream regulation to waste disposal, which impinged on water quality. With ads showing a giant bulldozer claw about to devour an idyllic rural landscape, the film succeeded in drawing public attention to environmental problems then facing industrial societies. In a memorandum to the North Rhine–Westphalian cultural ministry, one local conservationist estimated that in a three-year period the film reached nearly 1.5 million viewers, including audiences in West Germany, Austria, Switzerland, and the United States.[72]

Promotional materials for instructors in postsecondary schools demonstrated nature conservationists' desire to showcase their broader environmental agenda: "Nature protection today is more than the so-called 'protection of natural monuments' of the turn-of-the-century," noted one ad. In the postwar era, it "aimed not only at single animals, individual plants, or the protection of their habitat. The work of nature protection today is aimed more generally at the undamaged preservation of the bases of our existence, namely the water, soil, and climate. A landscape in which the best living conditions for plants, animals, and humans are available will in the long term be simultaneously outwardly beautiful . . . and an expression of biological health. . . . it will always be a true *Heimat.*"[73] Outward aesthetic beauty and ecological integrity thus worked hand in hand to expand the meaning of *Heimat* to include threats to water, air, and soil, and to wed homeland sentiment to future-oriented landscape planning.

Despite the expansion of nature conservationists' environmental vision, the references to *Heimat* in *Nature in Danger* revealed that important moral elements and patterns of cultural pessimism resonated across the 1945 divide. As Reinhard Demoll claimed in his work *Heimat: Bearbeitung und Gestaltung* (1958), only a renewal of ties to the homeland could serve as a stepping stone to international environmental awareness. Far from being antiquated, *Heimat* sentiment was the social basis of international cooperation in a postwar world; immersion in a healthy natural environment would enable West Germans to rebuild the sense of family, the ties of friendship, and an ethical relationship to the community. Without *Heimat* feeling, Demoll continued, the atomization and "massification" of West Germany would ensue, leaving its "uprooted" individuals prone to "impatience, intolerance, [and] rash judgments"; swaying between "spontaneous heroism today, soulless brutality and barbarism tomorrow"; and receptive "to every suggestion and therefore a willing victim of loud propaganda."[74] Like other postwar conservatives, Demoll envisioned *Heimat* as a bulwark against totalitarianism. In tandem with *Heimat*'s reinvention as a moral terrain untainted by fascism and impervious to materialism, environmentalists refashioned the homeland as the locus of ecological stewardship and conservative reintegration.

Demoll's adoption of *Heimat* ideals to promote environmental stewardship and counteract the leveling tendencies of mass society underscores the ideological connections between prewar traditions of environmental *Kulturkritik* and postwar environmental consciousness.[75] Like many academic, business, and government elites in the era of Chancellor Konrad Adenauer, Demoll blamed the debacle of Nazism on the uprooted "mass man" (as conceived by such thinkers as José Ortega y Gasset) rather than on the failings of individual political leaders, conservative and right-wing anti-republicanism, or the economic and social crises of the interwar years.[76] This anxiety about the masses, which had its roots in fin-de-siècle cultural pessimism, also reflected conservative explanations for the growing specter of communism, and it inflected the new ecological vision according to traditionalist social ideals and bureaucratic prerogatives. Nature conservationists, along with their allies in government, scientific societies, and professional organizations, attributed environmental degradation to cultural attitudes and a spiritual malaise rather than to structural contradictions within capitalist society itself. They targeted especially the "Americanized" consumer culture of the postwar decades for feeding West Germans' and other Europeans' "obsession with and addiction to money" and devastating the natural environment in the process.[77] And, rather than encouraging democratic, grass-roots environmental mobilization, anxieties about mass society tended to highlight the role of well-trained experts in managing natural resources, including the Rhine's watershed.[78]

The moral concerns that underlay the environmental rediscovery of

Heimat in the 1950s and 1960s found expression in references to Rhine pollution as a "disease of civilization." In a VDG article of 1956 entitled "Man and Nature," for example, one government minister described nature as the "fertile soil" of our "spiritual capacity for work and ability to exist." He called for a new religion of "deepened reverence for nature" that recognized the "infinite world spirit." Without such a belief, the author maintained, there could be no "culture" (*Kultur*) to contain the forces of "technology and civilization" (*Zivilisation*) and to purify the "dirty underside of the [economic] miracle."[79] Such references to a spiritual, nature-loving *Kultur* shaping the forces of mechanistic *Zivilisation* was another rhetorical distinction that stemmed from turn-of-the-century cultural pessimism. One North Rhine–Westphalian official in 1956 wrote that classical civilizations had often fallen as a result of deforestation or exhaustion of the soil, but modern industrial civilization would collapse as a result of dwindling and polluted water supplies. He argued that a sense of *Kultur* was Germans' only hope for rallying against "wastewater sinners" who dumped their refuse into the nation's waterways.[80]

The ability to take control of the Rhine's industrial wastes and render them environmentally harmless thus promoted a notion of the FRG as an ideal community of ethically responsible neighbors rather than a society of atomized, materialistic individuals. In the article "The Preservation of Federal Waterways," for example, the Düsseldorf water engineer Horst Klosterkemper suggested in 1960 that waste management and political restructuring went hand in hand. Klosterkemper's essay detailed the state's efforts to offer municipal grants to improve wastewater treatment plants and to assist industrial facilities in eliminating toxins through loan programs and tax incentives. Such measures would need to be expanded, according to Klosterkemper, so as to reduce the "burden of guilt" for pollution caused by ten years of war (1914–1918, 1939–1945), nine years of economic disorder (1918–1924, 1945–1948), and the period of the so-called Four-Year Plan (1935–1939)— "twenty-three years, therefore, in which people neglected or completely ignored the condition of our rivers."[81] He proposed that political instability, war, and economic disruption had prevented technical experts from managing the river's resources in a proper fashion. While municipalities had begun to install up-to-date treatment facilities between the wars, Klosterkemper continued, the National Socialist regime had prioritized rearmament and settlement projects over water purification, setting the stage for the water pollution crisis of the 1950s. Mary Douglas's analysis of pollution elimination as a creative act of environmental organization reflecting anxieties about changes in social and cultural norms has particular relevance for German water pollution control in the first two decades after the war. For many conservationists, hydrologists, and state officials, pollution control was not merely a technical solution to wastewater problems but a symbol of deeper processes of political atonement and cultural reconstruction.[82]

The reinvention of *Heimat* as a pathway to international cooperation also led water experts to recast the chauvinistic meaning of the *Kultur–Zivilisation* binary to correspond with an era of European integration on economic, cultural, and environmental matters. Before World War II, tensions between Germany and France had been symbolized in the opposition between *Kultur* and *Zivilisation*. In the occupied Rhineland after World War I, for example, German cultural officials had often portrayed French occupation forces as an alien civilization bent on destroying *völkisch* culture. The National Socialists had for propaganda purposes heightened the nationalist significance of this distinction. In an exhibition in 1941 titled "The Rhine, Germany's Eternal River," for example, organizers noted that the river was "Germany's river, but never Germany's border," thus celebrating the anti-French sentiments of Ernst Moritz Arndt's famous treatise, written in 1813 during the Wars of Liberation from Napoleonic hegemony.[83] Postwar references to *Kultur* versus *Zivilisation*, by contrast, expunged the anti-French rhetoric, instead describing nature as the source of the "eternal world spirit" and its destruction as a sacrilege against the divine forces of the cosmos. Highlighting the ethical malaise that perpetuated environmental destruction throughout Europe, this broadening of *Kultur* recognized that progress on the Rhine would ultimately require international solutions.

The 1950s proved to be a particularly opportune time to seek international cooperation on environmental matters. Collective security within NATO and economic and cultural integration through the European Coal and Steel Community provided the institutional frameworks for ameliorating nationalist conflicts, containing potential German aggression, and creating transnational regional networks for promoting economic development along the Rhine.[84] Cultural representations of the Rhine mirrored this change in political and cultural priorities. On a speedboat tour of the river in November 1945, Charles de Gaulle turned toward its eastern bank and observed, "the bank of the Rhine—yesterday still a line of battle, tomorrow the bond of union."[85] And in his treatise *Bridges over the Rhine*, Ernst Noth suggested in 1947 that Germans and the French could replace the bridges destroyed by the war with new structures that created cooperative pathways, rather than militarized access points, between the two countries.[86] The postwar Rhine was no longer "Germany's river" but an international waterway connecting European cultures and promoting economic development.

International efforts to restore the Rhine relied on its cultural reinvention as the first "European" river. As a program that transcended national boundaries, the Rhine clean-up effort stood as a model for the possibilities of European integration.[87] The Rhine Commission was one positive result of this desire for international cooperation; so, too, were consortia such as the International Working Group of the Waterworks of the Rhine Basin (IAWR), founded in 1970, which contributed to the development of collective wastewater treatment within the Rhine watershed.[88] As the Swiss engineer Otto Jaag noted in 1956, water problems demand "broad-thinking.... when several are dependent on the same waterway, there can be no one-sided thinking in the long run."[89] As Jaag also stated, "There must be a high degree of understanding and unselfish willingness to assist in such international tasks in order to make the efforts effective."[90] Here again, scientific knowledge became a vehicle for forging cross-national alliances. Reporting on the Rhine Commission meeting in Meersburg in 1956, the VDG noted that "biological thinking and openness were evident," suggesting that cooperation grounded on ecological thinking, rather than irrational nationalism, would solve the Rhine's pollution problems. As a spokesman for the European Water Protection Federation remarked, "It shouldn't be the case that we gain the power of the atom but at the same time lose the vital element water!"[91]

Despite such pronouncements of international goodwill and the development of an ecosystemic understanding of the river's degradation, the Rhine Commission and the International Working Group did not achieve effective pollution control measures during the first three decades after the war. National interests quickly reasserted themselves, particularly in negotiations on the dumping of chloride salts into the Rhine. Scientific analyses sponsored by the commission determined that one Alsatian potash mining company accounted for nearly 40 percent of the chloride pollutants in the Rhine. But France's refusal to pay for an alternative disposal method (injection of the salts deep into the earth) caused so much animosity among the member states that in 1979 the Dutch took the unprecedented step of recalling their ambassador from France for consultation.[92] Conventions proposed to clean up chemical and thermal pollution also came to naught, largely because the technologies needed to eliminate certain substances in treatment plants were either too costly or did not yet exist. The highly formal "command-and-control" style of regulation favored by the commission tended to thwart international cooperation because it was often out of sync with the varying regulatory cultures of the five member states.[93] To cite one example, weaknesses in Germany's Water Management Law of 1957 conferred responsibility for water resources on states, which fought hard to retain their independence in regulating the discharge of effluents, establishing drinking water standards, and negotiating with non–European Community member Switzerland.[94]

The Sandoz chemical fire in November 1986 proved decisive in surmounting this international regulatory impasse. In this incident, the so-called "Bhopal on the Rhine," water used to extinguish the fire washed thirty tons of pesticides, fungicides, dyes, and other toxic chemicals into the Rhine; half a million fish died instantly. The incident prompted widespread demonstrations and government pressure, spurring the Rhine Commission to draft a new set of pollution control standards that became known as the Rhine Action Plan. Based on a more flexible model of regulation that allowed states and private industry to devise their own methods for achieving shared environmental goals, the plan dramatically reduced effluents in the river.[95] It also committed West Germany, the Netherlands, France, Switzerland, and Luxembourg to ensuring the Rhine's continued use as a source of drinking water, to purifying sediment that had become polluted by heavy metals and toxic chemicals, and even to restoring the main channel of the river as a habitat for salmon and other migratory fish.[96] In order to safeguard the Rhine as a source of drinking water, the five governments pledged to cut the discharge of "blacklisted" substances by 50 percent by 1995, at enormous expense; according to one estimate, the German chemical industry alone spent DM 6.6 billion on sewage treatment between 1987 and 1991. A Rhine Commission study released in 1989, *Ecological Master Plan for the Rhine*, also known as *Salmon 2000*, laid the groundwork for some of the most ambitious ecological restoration work ever undertaken, such as the regeneration of alluvial floodplains, the restoration of spawning beds, and the construction of fish ladders around weirs and dams.[97]

The plan's focus on ecosystem management, pollution control, and habitat restoration was a far cry from the *Heimatschutz*-inspired aesthetic, ethical, and nationalist ideals that had informed the Nazi landscape protection plan of 1937 to 1941. In the intervening decades state officials, environmental organizations, scientists, and hydrologists had shifted the goal of environmental reform from maintaining the landscape's visual harmony to cleaning up and restoring ecological systems. More than any other environmental issue facing postwar Germany, water resource management in general, and the status of the Rhine in particular, provided the context in which direct "environmental management" of toxic effluents could emerge in the 1970s as

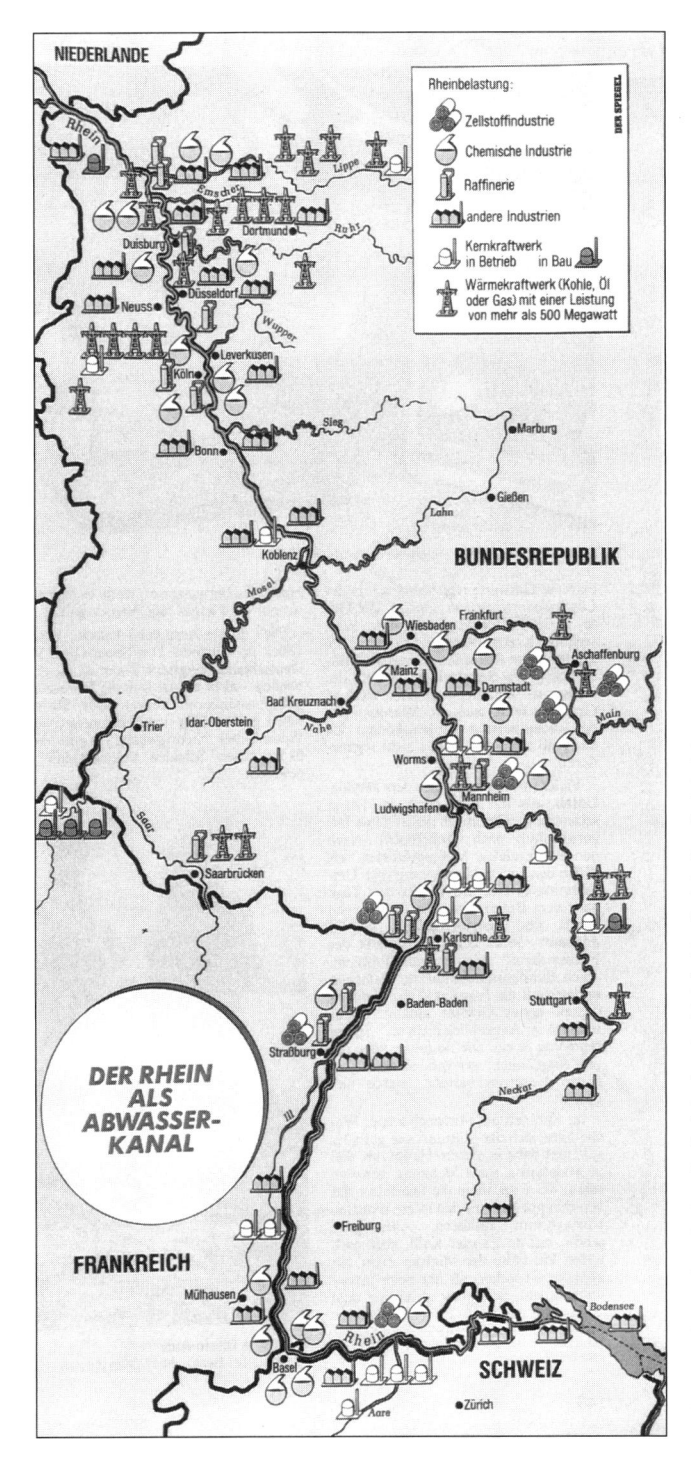

Figure 8.1. "The Rhine as Wastewater Canal." This map shows the sites of pulp factories, chemical plants, refineries, nuclear power plants (either in operation or under construction), and thermal power plants with a capacity of greater than 500 megawatts along the banks of the Rhine and its main tributaries from the Netherlands to Switzerland. Reproduced by permission from *Der Spiegel* 47 (1986).

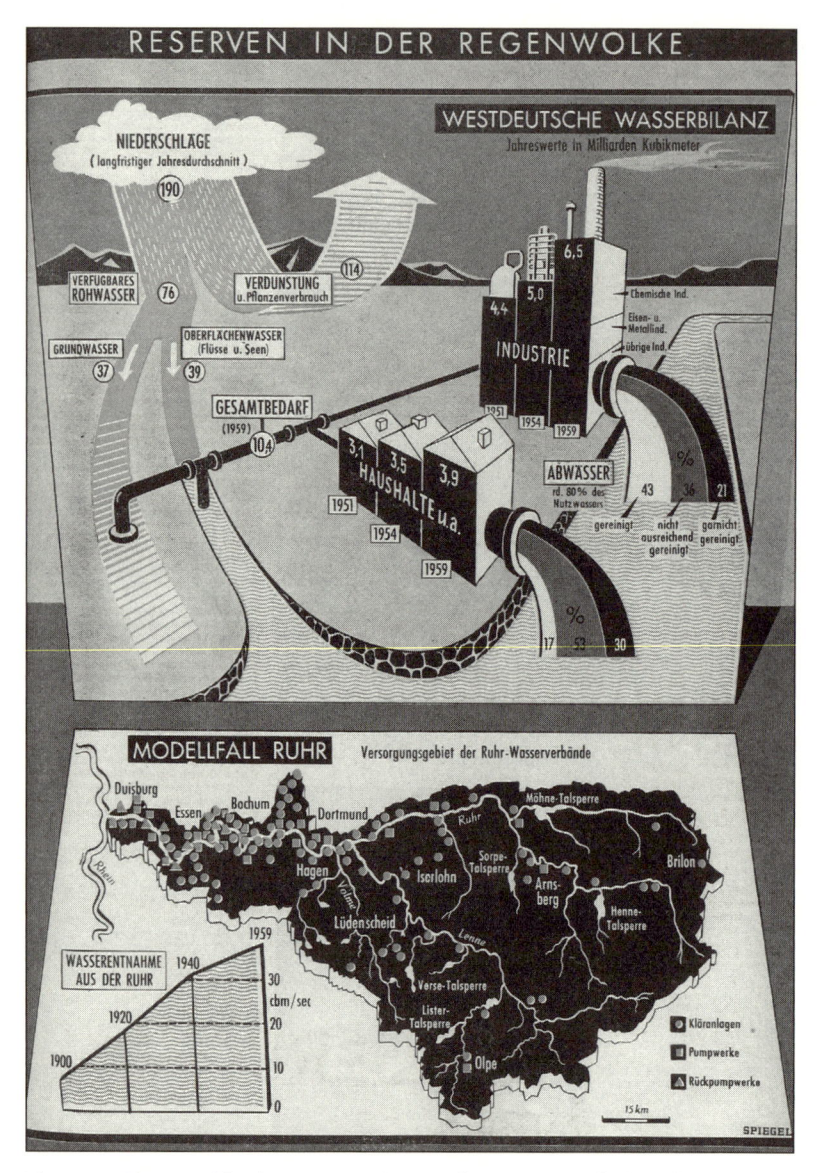

Figure 8.2. Two graphics from a magazine article on Germany's dwindling and increasingly polluted waterways. The top graphic represents the water cycle in West Germany, while the lower one shows the location of water treatment plants and pumping stations along the rivers in the Ruhr Valley, formerly one of Germany's most densely populated and heavily industrialized regions. Reproduced by permission from *Der Spiegel* 47 (1986).

a field of political engagement at the domestic and international levels.[98] Even though the riparian states could not achieve consensus on how to address these problems before the mid-1980s, they did begin to see the Rhine during these years as a fragile life-support system rather than merely a shipping corridor or enormous septic tank. Scientific evidence about the type and extent of pollutants, their potential health effects, and their ability to contaminate drinking supplies and kill fish far from their origin revealed the inadequacy of setting aside scenic landscapes and urban green spaces as an environmental strategy. Instead, conservationists, engineers, and scientists turned toward a nascent model of ecosystem management that strove to purify and maintain the human habitat through regulatory permits and control technologies.

The turn toward ecological perception served broader political and cultural purposes as well. In West Germany, pollution concerns dovetailed with conservationists' efforts to redefine *Heimat* as a locus of positive emotional identification, to disseminate conservative social values, and to facilitate international cooperation in a region still traumatized by the effects of war. By recasting the distinction between *Kultur* and *Zivilisation* in terms of a spiritual malaise afflicting all of European society, moreover, environmental advocates helped to reinvent the Rhine as an intricate, vulnerable organism whose arteries transcended national boundaries. Once a catalyst for the economic rivalry and nationalist hatred that led to World War II, the Rhine came to symbolize a new European era of "postmaterial" ecological values and transnational cooperation and integration.[99] The resulting environmental awareness was an uneasy amalgam that embraced both postwar ecological analysis and prewar cultural criticism, resulting in a conservative form of environmentalism that addressed pollution and other environmental threats through technocratic management rather than sustained grassroots mobilization.[100]

The continued vitality of these conservative impulses within the Federal Republic's conservation, scientific, and engineering communities should caution us against seeing the environmentalism of the 1950s and 1960s as a stepping stone to 1970s green activism or marking a linear path from *Heimat* to *Umwelt*.[101] Although the emergence of the Greens lies beyond the scope of this chapter, it would be worth investigating whether the movement's true innovation was not so much in introducing environmental problems to the German public but rather in mobilizing young, disaffected voters to pursue a variety of socially progressive aims under the umbrella of ecology. Through grassroots tactics and an appeal to direct democracy, the citizens' initiatives and party lists that eventually coalesced into the Green Party offered an alternative to the conservative and bureaucratized forms of nature conservation that dominated the 1950s and 1960s and, as the Wyhl controversy and Sandoz spill demonstrated, failed to meet the challenge of nuclear and chemical contamination of the Rhine.[102] In fact, conservative environmentalists such as Bernhard Grzimek and Konrad Lorenz were initially attracted to the Green movement's ecological focus and abandoned the party only when it began to adopt leftist policies on social and diplomatic issues, such as women's rights, nuclear disarmament, and pacifism.[103] After the 1970s, civil servants within Germany's forestry and natural resource conservation ministries carried on bureaucratic forms of environmental protection with antecedents in the 1950s. At the same time private organizations, such as the Rhenish Society for Landscape Protection and Monument Conservation, reshaped earlier *Heimatschutz* and landscape protection agendas for an age of mass tourism and automobility. The Green Party's leftist environmental alternative has therefore never entirely monopolized ecological attention or displaced conservative and even right-wing ecological concerns, a factor that perhaps explains the depth of environmental commitment, if not political agreement, among today's German citizens.

9

Postwar Perceptions of German Rivers
A Study of the Lech as Energy Source, Nature Preserve, and Tourist Attraction

UTE HASENÖHRL

Water in any form—whether ocean, lake, or river—has always been a major tourist attraction. Rivers, especially those with rapids, are often perceived as symbols of untamed nature. Sports enthusiasts, hikers, and nature lovers choose rivers to savor peace, tranquility, and landscapes unspoiled by development. Whereas this kind of tourism is typical for individual vacationers, cruises and package holidays designed for mass audiences often concentrate on specific parts of a river. They focus less on rivers' ecological and aesthetic qualities and more on their suitability for recreation and entertainment. For instance, scenic locations are either made accessible or created (as in the case of artificial lakes or reservoirs), or the river and its surroundings are developed for swimming, boating, and other water sports. In tourism, two views of rivers—as sublime natural landscape and as recreational facility—both collide and overlap.

Tensions exist not only between these different types of tourism but also between tourism and other interests, such as industry, energy production, fisheries, and nature conservation, leading to changing alliances in competitive situations. For example, in clashes over the construction of hydroelectric power plants and the regulation of rivers, both conservationists and energy companies have sought to gain public and local support by arguing that their respective visions of a river (natural versus designed) would most benefit the development of tourism.

By examining the relationships between nature conservation, energy production, and tourism in conflicts over hydroelectric energy, this chapter addresses the following questions: In which disputes was tourism an important factor? What was its significance in the protagonists' strategic approach? Which kind of tourism was being encouraged? Did the protagonists' strategy help to win the support of the local population? Were rivers in particular and nature in general perceived as possessing inherent aesthetic value? In this regard, it is striking that developers as well as their opponents drew on Romantic imagery of Germany's rivers to support their respective positions. How conflicting views managed to appeal to the same aesthetic conventions is thus one of the most important questions to be addressed in this study. The perspectives of the tourism industry and of individual tourists can, however, only be touched on here.

My analyses are based on a case study of the Lech River during the second half of the twentieth century. While the findings are representative of similar conflicts regarding other pre-Alpine rivers and waterfalls in this time period, they may not be valid for other types of waterways.[1] Pre-Alpine rivers differ from, for example, large streams because they are relatively unsuitable for transport due to their location in mountainous terrain, as well as the mostly agrarian and comparatively sparsely populated character of the adjoining area, especially upstream. In the following discussion, *conservationists* is used as a collective term that includes both civil servants (employed by state and local governments to protect publicly owned natural resources) and environmentalists (affiliated with nongovernmental conservation organizations such as the Bund Naturschutz in Bayern and the Deutsche Alpenverein).

Constructing Hydroelectric Power Plants in Bavaria

In Bavaria the earliest plans to systematically utilize pre-Alpine rivers for energy production date back to the late nineteenth century, when industrialization steadily increased the demand for power. Technical innovations, such as the invention of generators and the introduction of high-voltage lines, soon enabled production of electricity from flowing water as well as its transmission to distant places.[2] A memorandum of the Königliche Oberste Baubehörde (Royal Planning Commission) dated 1907 therefore declared the use of rivers for hydroelectric energy production to be of the highest priority.[3]

A river's capacity to produce energy corresponds to the volume of water it contains and the height from which it descends. Because of their steepness and velocity, pre-Alpine rivers are well suited for generating power, even if their catchment areas happen to be limited.[4] While demand for power is highest in winter, the flow rates of pre-Alpine rivers peak between April and September after winter snows melt. Thus, as early as 1926 the Oberste Baubehörde proposed combining hydroelectric power plants with reservoirs to ensure an adequate water supply during periods of high demand.[5]

The need for hydroelectric power grew dramatically after World War II. Among the West German federal states, Bavaria's difficulties in meeting the energy needs of its people were the most severe. In 1946, Bavarian power plants were able to generate only about 60 percent of the amount of energy that had been available in the years leading up to the war. To make things worse, war refugees had increased the state's population by almost a third. Thus Bavaria—with few coal deposits of its own and struggling to obtain additional coal from the Ruhr basin after imports from the lignite regions in East Germany had been cut off—suffered a drastic decline in industrial production due to the rationing of electricity.[6] In the winter of 1948–1949 alone the loss was equivalent to DM 250 million. At times, even essential needs such as energy for refrigerating warehouses and dairies, could not be met. To resolve this crisis and to make Bavaria less dependent on fuel and electricity imports, the state initiated a vast energy program based mostly on hydroelectric power.[7]

Acknowledging the urgent demand for energy during the postwar period, conservationists tolerated these activities.[8] They promoted the concentration of hydroelectric power production on projects with considerable value for the general energy supply, provided that installations were integrated into the landscape aesthetically. At the same time, they fiercely opposed those facilities of only local or regional importance in areas of outstanding natural beauty. Water power, they claimed, should be combined with other energy sources like steam or nuclear power.[9] Indeed, the federal state of Bavaria soon abolished the notion of self-sufficiency and the one-sided emphasis on water power but without giving up its plans for additional hydroelectric power plants.[10] However, after the immediate power problems had been solved in the early 1950s, especially after the nuclear conference in Geneva in 1955 with its promise of nuclear energy as a potentially unlimited power resource, conservationists regarded the use of water power as merely an interim solution.[11]

Power Plants on the Lech

Between 1940 and 1984 the Lech, which flows from Austria into southern Bavaria, was transformed from a scenic waterway into an important source of hydroelectric power. Various rhetorical strategies were used by both conservationists and the energy company Bayerische Wasserkraftwerke AG, widely known as BAWAG, to present their respective viewpoints to the local population, the wider public, and the state authorities. The dispute between BAWAG and environmental activists, while fiercer and more protracted than usual, can nevertheless be regarded as representative of conflicts over the regulation of Bavarian rivers, in which the modes of argumentation were largely the same.[12] Energy company officials emphasized economic growth, technical problems with conservationists' proposed solutions, and the binding character of existing license agreements with the state. Conservationists stressed the unique beauty of the landscape; the necessity of protecting threatened species; potential alternatives, such as other construction sites or energy sources; as well as the priority of the environment and the homeland over short-term profits. Both sides repeatedly raised the possibility of developing tourism in these scenic areas.

Conservationists presented aesthetic, ethical, and ecological arguments and invoked legal, economic, and political ones as well.[13] Focal points, however, varied according to the situation and audience addressed. The beauty of the landscape and the obligation to preserve at least remnants of the shrinking habitat of unregulated rivers for future generations were almost always cited.[14] Environmental considerations in general played only a minor role and were mostly utilized to support conservationists' contention that BAWAG should concentrate primarily on the remediation of the Lech downstream, which had already been significantly altered by earlier engineering projects.[15] In such cases dams were regarded as ecologically beneficial because of their stabilizing effect on the groundwater table.[16] The broad attention given to legal and political aspects was a response to BAWAG's questionable conduct. Being sure that its requests would be granted with almost no restrictions and that it would not be called to account for breaches of law because of its status as a partly state-owned company, BAWAG repeatedly tried to create a fait accompli in the 1950s and 1960s.[17] Conservationists argued that this kind of power politics would not only render conservation itself pointless and serve as an example for future transgressions, it would also undermine the democratic process.[18] They tried to prove that destruction of the last unregulated river sections served only to increase BAWAG's profits and was detrimental to the public interest. The potential power production would be negligible for the overall energy supply, even more so since steam and, later, nuclear power were much more efficient.[19]

BAWAG, on the other hand, cited either the general planning of 1940, its contract with the state, or geological and economic necessities. The company's core argument revolved around its contract in 1949 with the Bavarian government, which had granted BAWAG the exclusive right to utilize the entire Lech for power production as well as the right to operate all its waterworks for the next ninety years.[20] The company asserted that its primary aim was not to maximize profits but to secure Bavaria's power supply, thus fulfilling a state mandate and serving the public interest.[21]

In their overall arguments, both conservationists and BAWAG stressed energy production, political factors, and ecological concerns. Another important aspect frequently referred to by both sides was the potential for tourism and recreation, either on the undeveloped Lech or on future reservoirs. However, as Frank Uekötter pointed out in his analysis of the arguments used to prevent the building of a racetrack in Germany's Sauerland (in North Rhine–Westphalia), conservationists were rather unclear about just how pristine nature or a traditional cultural landscape would contribute to the development of tourism. Instead, their statements consisted largely of vague promises and even the occasional cliché. As in the case of the Lech, the idea of tourism was often used to strengthen a broader argument.[22] In such instances it was usually combined with aesthetic and economic justifications or associated with the public interest, which both groups claimed to defend.

Public interest was defined on the one hand as access to sufficient amounts of affordable electricity, resulting in economic growth, prosperity, and a high standard of living.[23] On the other hand, it was also understood to include the preservation of landscapes of outstanding natural beauty and a healthy environment for current and future generations.[24] The predominant strategy of using tourism as a rhetorical weapon was, with good cause, closely linked to the concept of public interest, which itself was grounded on the fact that larger rivers, as public waterways, are common property and therefore subject to state control. Thus they cannot be used arbitrarily for private gain. In fact, interested parties must legally demonstrate either that their actions will not impair the claims of others or that their goals coincide with and will significantly advance the public interest.[25] Arguments in favor of tourism and recreation inevitably linked the economic self-interest of the local population with principal concerns of the public as potential vacationers in need of attractive destinations and recreational resources.

Conservationists argued that dams would undermine the development of tourism, since the distinctive beauty of a natural landscape would always prove more attractive than artificial lakes, which were periodically surrounded by expanses of mud caused by fluctuations in the reservoirs' water levels.[26] Furthermore, they cited the public's constitutional right to enjoy nature and outdoor recreation, which gained importance as leisure time increased and which could likely be impaired by the destruction of scenic areas.[27] BAWAG, on the other hand, asserted that dams and artificial lakes could not only be integrated harmoniously into the landscape but, in the case of pre-Alpine rivers, would even increase their beauty, thus having a positive effect on tourism.[28] The claim to be able to create lakes and landscapes even more beautiful than naturally occurring ones was regarded by conservationists as both presumptuous and insincere.[29]

Mass Tourism versus Green Tourism

Conservationists and energy companies represented two contrary concepts of tourism and recreation. While conservationists focused on individual travelers looking for peace and solitude, energy companies pointed out the opportunities provided by reservoirs for water sports and mass tourism. The relationship between nature conservation and tourism was ambivalent. On the one hand, the success of nature protection campaigns often corresponded to the potential that a natural or cultural landscape offered for tourism and, therefore, to the economic opportunities it would provide for local residents. On the other hand, there were clear tensions. While conservationists, in focusing on a wild and unspoiled nature, were at least theoretically in harmony with individual tourism, they objected strongly to mass tourism and the hazards posed by crowds of tourists in fragile environments, by development of infrastructure to meet their needs, and by efforts to attract them by redesigning the landscape.[30] Accordingly, conservationists contrasted individual and mass tourism by countering the positive term *recreation* with the dismissive image of an amusement park (*Rummelplatz*).[31]

Even though conservationists' notions of a nature-oriented tourism did not coalesce into a coherent concept during the 1950s and 1960s, the first signs of green tourism can be identified. Criticism of tourism's negative effects is as old as the phenomenon itself.[32] Moreover, alternative forms of traveling, such as the hiking trips organized by Touristenverein Die Naturfreunde (the traveling, hiking, and nature conservation club of the German labor movement) or the Deutscher Alpenverein, date back to the late nineteenth century.[33] Nevertheless, it was not until the late 1970s that discussion of the sociocultural effects of tourism on developing countries and concern about the destruction of the Alpine landscape led to the idea of green tourism, now also known as ecotourism.[34] This type of tourism aims to minimize visitors' impact on the ecological, social, and cultural environments of a holiday destination. Instead of creating an artificial tourist infrastructure, which would alter and possibly damage the site, existing facilities should be utilized as much as possible. Only changes beneficial to both tourists and the local population and compatible with the ecosystem are permissible. To curtail the negative effects of mass tourism, the number of visitors may be limited and access to sensitive habitats restricted.[35]

Similar proposals had, however, already appeared in earlier discussions about the construction of cable cars in the Bavarian Alps.[36] They had also surfaced time and again in conflicts over the building of hydroelectric power plants for electricity production. Central to the conservationists' philosophy was the belief that as industrialization and urbanization increased—generating more and more noise, air pollution, and crowds—humans' and especially urban dwellers' need for pristine nature, harmonious landscapes, peace, and solitude would increase as well.[37] Communities that kept their surroundings free of attractions designed to draw mass audiences would profit in the long run by resisting development, especially since places where nature remained intact were becoming rarer and therefore more valuable.[38] The conservationists also pointed out that facilities like cable cars or bodies of water developed for recreational uses were frequented mainly by day-trippers, who tended to scare off long-term vacationers. Moreover, nature-loving hikers' or mountaineers' penchant for exploring wilderness should be considered just as much as the desires of tourists for entertainment and artificial attractions.[39] To reduce their impact on scenic areas and especially on nature preserves, visitors could be tactfully directed to less environmentally sensitive zones.[40]

By contrast, energy companies promoted the economic potential of mass tourism in a landscape transformed by their activities.[41] The appeal of the rivers and their communities would grow due to new attractions such as reservoirs, which could be used for water sports and local recreation. Their beautiful appearance would also lure motorists who would otherwise have simply passed by on their way to another destination.[42]

Participants in the debate assumed that their respective notions of tourism would be shared by the public, or at least by a considerable part of it. Yet conservationists' vision of nature-oriented tourism was frequently regarded by the local population as quixotic and inefficient in the face of growing competition among tourist areas. Residents therefore tended to agree with the energy companies that features such as reservoirs would produce greater benefits for the regional economy. Catering exclusively to nature lovers would also mean smaller profits than might be realized from mass tourism.[43] Vacationers also seemed to readily accept reservoirs as destinations for day trips or holidays, provided their expectations were not shattered by ugly layers of debris or mud around the shoreline, as was the case at the Forggensee for several years in the 1950s and 1960s.[44]

The Tourist Gaze

The respective notions of tourism advanced by conservationists and energy companies were strongly influenced by differing conceptions of nature. Conservationists valued the uniqueness and authenticity of the undeveloped landscape, which in their view would always prove more attractive than artificial substitutes such as regulated rivers or reservoirs. The energy companies, for their part, preferred images of manicured riverscapes and lake resorts to the apparent disorder of unruly pre-Alpine rivers. On the surface these divergent ideas of nature and tourism correspond to the two main varieties of "the tourist gaze" identified by John Urry: the "romantic" and the "collective."[45] In the romantic form of the tourist gaze, the emphasis is on solitude, privacy, and a semi-spiritual relationship with the object of the gaze, mostly "undisturbed natural beauty." While for Urry the romantic gaze is largely a middle-class phenomenon, he regards the collective gaze, with its focus on conviviality, sociability, and shared experience, as essentially working class.[46] When looked at more closely, however, the respective attitudes of conservationists and energy companies toward nature cannot simply be equated with the romantic and the collective tourist gazes.

At first glance, the conservationists' image of nature appears to match the concept of the romantic tourist gaze, as this group promoted an experience of peace and solitude in a pristine landscape. But, in the case of pre-Alpine rivers especially, the landscapes that conservationists attempted to preserve did not always conform to the romanticized views of nature developed and diffused during the nineteenth century, when visual tropes and aesthetic ideas about nature were evolving in response to the enormous changes caused by the Industrial Revolution. Landscape paintings were, for instance, no longer limited to the idealized Arcadian landscape of the Roman Campagna.[47] Instead, the motifs popularized by Romantic painters, such as mountain peaks, turbulent seas, misty ruins, or stormy night skies, were meant to inspire sublime emotions in the viewer as he or she beheld an awesome nature animated by divine spirits.[48] In addition, greater opportunities to travel and a quest for national identity in the wake of the Napoleonic Wars led to the discovery of regional landscapes by painters of the early nineteenth century.[49] The Biedermeier art of the Munich School thus turned toward the Upper Bavarian landscape, following a realistic plein-air approach and aiming to represent landscape neither as a space free of people nor as a metaphysical metaphor but as a topographically exact living space. While these paintings often glorified the homeland as a clean, sunny, ideal world, such leading painters as Eduard Schleich, Christian Morgenstern, Adolf Lier, and members of the Leibl School concentrated on transient and subjective aspects of the landscape, especially color, light, and atmosphere.[50]

Widely available in inexpensive prints and frequently reproduced in magazines, the Romantic- and Biedermeier-era landscape paintings were critical in shaping popular perceptions of nature. By the end of the nineteenth century these images had become imprinted on the collective memory as a set of familiar stereotypes and visual clichés that typified the homeland and evoked a stable and balanced relationship between humans and nature. They thus helped to create expectations of how a "typical" Upper Bavarian landscape in particular and an "attractive" landscape in general should look. Because tourists respond strongly to contemporary images of their (potential) destinations, these "place myths" played a crucial and formative role in how tourism developed.[51]

Biedermeier landscape artists painted pre-Alpine rivers as well, especially the part of the Isar that lies south of Munich. This area was frequented by the Munich bourgeoisie on holidays and became a popular motif in paintings they bought to beautify their homes.[52] In general, however, pre-Alpine rivers were romanticized, though sometimes the river's atmosphere rather than the river itself dominated the painting.[53] Pre-Alpine rivers often form broad runnels in a barren landscape, with gravel banks and many small, rocky islands. Along these rivers only pioneer societies can thrive because seasonal floods continually redefine the riverbed.[54] This rugged character fits neither the Romantic stereotype of a white-water river rushing through a narrow gorge nor that of a majestic waterway meandering through a valley covered with abundant vegetation and crowned by an ancient castle or chapel.[55]

Energy companies did not concentrate just on the recreational potential of reservoirs, as representatives of the "collective tourist gaze" might be expected to do, but also promoted the aesthetic benefits of the lakes they attempted to create.[56] Furthermore, the authenticity of the new formations—an issue of special importance to the romantic tourist gaze—was accentuated wherever possible, and company representatives asserted that with the construction of reservoirs, glacial lakes were being replicated.[57] The following passage from a BAWAG brochure on the newly created Forggensee at the

River Lech exemplifies this approach: "[T]his vivid presentation shows the harmonious integration of the new Alpine lake into the landscape. Everyone attracted by the beauty of the foothills of the Alps will warmly welcome this enhancement to the string of lakes near Füssen. Indeed, the stored water of the Lech partially refills a glacial basin, resulting in a natural integration of the reservoir into the surrounding landscape."[58] Thus, while conservationists defended a landscape that shared little more with idealized Romantic images than the rough dynamics of raw nature, energy companies hoped to evoke Romantic associations with their claims of "glacial" authenticity and the image of idyllic mountain lakes.

Energy companies sought to exploit both possible tourist markets. First, they emphasized the opportunities for entertainment, recreation, and water sports that reservoirs offered. Second, they skillfully compared Romantic stereotypes of nature to the apparently dreary and barren character of pre-Alpine rivers: "Where formerly the River Lech carved its bed, resulting in a sinking groundwater table and unsightly debris fields, a string of lovely lakes created by the barrage was being generated. They form an example of how technical constructions can be integrated into natural surroundings almost completely through farsighted and careful planning."[59] Conservationists in turn exhibited most of the characteristics of the romantic tourist gaze, such as the yearning for peace and solitude in pristine nature. In the case of pre-Alpine rivers, however, the landscape they aimed to protect resembled idealized images of nature far less than the artificial reservoirs did. This un-Romantic appearance may be one reason why conservationists' descriptions of these rivers often remained vague.[60]

THE APPEAL of the natural or designed riverscape rested largely on how well it matched popular perceptions of nature. Contemporary newspaper and journal articles reveal how unpopular pre-Alpine river landscapes seem to have been in the 1950s and 1960s in comparison with reservoirs, especially the Forggensee. Not all writers were as enthusiastic as Wilhelm Jakob, who wrote, "[T]he great symphony of nature has not been disturbed; on the contrary it has flared up even more powerfully, as if basses of outstanding vigor and color had been added." But many articles did contend that dams and reservoirs did no harm to the beauty of the landscape.[61] Further research is still needed, however, to properly assess the degree to which the seemingly unaesthetic visual character of the Bavarian pre-Alpine rivers determined public support for their protection and whether or not this drab aspect played any role in conservationists' failure to preserve more natural landscapes.

Generally speaking, while some areas of outstanding natural beauty or natural monuments could be preserved, the vast majority of rivers had been regulated and, wherever possible, utilized for power production.[62] Still, it cannot be merely coincidental that in regard to pre-Alpine rivers, nature protection seems to have been most successful in the case of romanticized features such as waterfalls and gorges. The same connection seems to apply in the case of the Isar, which is the one pre-Alpine river prominent in Romantic and Biedermeier paintings and which also happened to be a popular recreation area close to Munich.[63] Thus, to a certain degree, conservationists' prospects of success did depend on how closely the riparian landscape they sought to preserve mirrored the visual stereotypes popularized by Romantic images of nature—a resemblance that was also directly connected with the economic potential of tourism in these areas. This was of crucial importance, since conservationists could not rely on the state to act as a mediator or ally. On the contrary, the state administration often supported energy companies in order to ensure Bavarians' present or future power supply.[64]

Still, conservationists' effectiveness also relied on a broad range of determining factors: the profitability and size of a project (and thus its value for the state's electricity supply), the specific time at which it was proposed, the political influence of the actors, the economic structure of the region, as well as the attitudes of the local population and the general public. For major or minor hydroelectric projects, the economic relevance of the power stations as producers of energy and as local employers frequently took priority.[65] Furthermore, minor projects often required only minimal intrusions on the environment. In these cases conservationists had to be, and normally were, content with concessions regarding design and integration of buildings and reservoirs into the landscape.[66]

Conservationists tended to be more successful with medium-sized projects. For example, proposals to construct hydroelectric power plants at Waginger Lake in 1949–1951, in the Werdenfels region in 1949–1950 and 1953–1954, and at the Partnach Gorge in 1948–1950 were prevented despite the dire energy crisis of the postwar years.[67] The plants would neither sufficiently increase power production nor create enough new jobs to justify the disruption and degradation of the landscape that were certain to result from these projects. Moreover, these riverscapes and natural monuments were often either major tourist attractions or important to local peoples' sense of identity. Thus the utility of the power plants for the local economy and the general energy supply was too small to justify their cost.[68] In contrast, an intact natural landscape might support a tourism industry that offered the prospect of prosperity for many area residents.[69] In circumstances such as these, alliances between conservationists and the local population often proved effective.

On the other hand, if the construction of reservoirs or other forms of development promised to attract even more tourists, the locals would favor the energy companies, especially in rural districts that otherwise depended on agricultural incomes.[70] Additionally, coalitions between conservationists and the local population were prone to fail when disputes arose about how rescued areas should be managed; in many cases conservationists wished to establish nature preserves, while residents preferred to develop these areas for tourism and recreational uses.[71] In the end, Romantic appearance or ecological significance alone saved only a few exceptional waterfalls and gorges—if these considerations really were decisive. Evidence indicates that if the state administration demanded and supported energy projects, vehement local or even national resistance could not stop them.[72]

10

Viewing the Gilded Age River

Photography and Tourism along the Wisconsin Dells

STEVEN HOELSCHER

Among the many uses to which America's rivers were put during the Gilded Age, tourism stands out as an especially vital bridge between that earlier era and our own. Nineteenth-century Americans called on rivers to transport raw materials, finished goods, people, and industrial waste; they dammed them at every turn to minimize the risks of flooding and to maximize both energy production and water diversion; and they used them to mark the boundaries of competing imperial systems. They began to envision rivers in other terms as well. Romantic philosophers and, later, middle-class Americans came to believe that waterways could be exploited not only for their utilitarian functions but also for their recreational and restorative potential. Rapid industrialization and urbanization in the late nineteenth century not only refashioned the physical environment of American cities but also profoundly affected how people perceived and interacted with the natural world outside the city, especially rivers. Long before the Colorado River became a mecca for white-water rafters, Victorian-era tourists plied the Hudson, Connecticut, and Wisconsin rivers seeking respite from the socioeconomic changes that were transforming their everyday lives.

This chapter examines the relationship of early tourism to one of those historic waterways—the Wisconsin—at a pivotal moment of social, economic, and environmental change. A half-day train ride from Chicago, the Dells of the Wisconsin River became the great inland metropolis's rural retreat during this period of massive social and geographic transition. That the river would become so connected to the city was hardly self-evident or inevitable. It was instead the consequence, to a considerable extent, of the power of the photographic view to construct an imaginative geography at once sublime and picturesque, both wild and morally uplifting. Indeed, long after the lumbering and settlement frontiers had passed, photographs of the river depicted a frontier playground where recreation and mass leisure replaced the challenge of "discovering" new land and where vacationing Victorians supplanted Native Americans as the inhabitants of a new American space. No longer simply a natural resource to exploit, the river became something to view. Photography, in short, turned a working river into a riverscape.

The photographic view, a distinct type of landscape photography that was already enormously popular by the start of the Civil War, reached the peak of its influence during the closing decades of the nineteenth century, when it became an important expression of national identity. "To see their country, to understand not just its topography but its underlying values and beliefs," Peter Bacon Hales writes, "nineteenth-century Americans increasingly turned to the photographic view."[1] While view photographers focused their cameras on a wide range of landscapes—rugged mountains, barren deserts, subtropical swamps, ghastly battlefields, monumental city buildings—among the most frequently pictured natural features were rivers. This is by no means an accident. From Niagara Falls at the beginning of the nineteenth century to Yellowstone at that century's end, rivers were routinely portrayed as representing the nation's "underlying values and beliefs."[2]

My argument about view photography's signal importance for America's waterways in the Gilded Age rests on three premises. First, as part of a more general "search for order" during a period of radical social and economic unrest, photographic views refracted an ideology of human control over nature in the creation of a new middle-class, postfrontier space.[3] Nature became picturesque scenery and serviceable for mass-produced inspiration. Second, nature's scenic transformation was succinctly accomplished by photography—a superb vehicle of cultural mythology. Because it uses a combination of technology and artifice to create a convincing illusion of reality—one perceived not as an image carefully shaped by the photographer's skill and sensibility but as a "factual" record—the medium possesses a unique power to naturalize social constructions. Complex ideologies of progress, development, and regional transformation may be presented as laws of nature, as incontrovertible facts. Third, photography's apparent transcription of reality proved an immensely valuable asset for one enterprise in particular: tourism. Acquiring photographs gives shape to travel as it informs what the viewer should see, how it should be seen, and when it should be seen—all in a matter-of-fact and seemingly "unmediated" way. These "pre-texts," as Joan Schwartz has argued, not only dictate the subject matter of travel photographs but also become one component of a cycle that unites itineraries, representations, and places. The resulting imaginative geographies are just that: new ways of seeing and experiencing social space that are themselves historically and geographically contingent and that are the product of human imagination rather than forces of nature.[4]

Examples of view photographers on America's rivers tangled in the web of art, commerce, tourism, and nature are well known to scholars of nineteenth-century landscape. The very best created river views that stunned American audiences. William Henry Jackson, Carleton Watkins, Timothy O'Sullivan, and F. Jay Haynes, to name only the most famous, composed and printed riverscapes that made their way into Victorian-era parlors, government reports, railroad stations, and museums across the country.[5] Their impressive photographs of sublime scenery throughout the Far West on such rivers as the Columbia, the Willamette, the Merced, and the Colorado set the standard for pictorial composition and technical competence.

Equally important, but working out of the considerably less grand midwestern landscape of the Wisconsin River Dells, Henry Hamilton Bennett was a contemporary of the frontier photographers (figure 10.1). His views of the river—especially his stereographic photographs and outsized panoramas—have been noted by art historians for their technical refinement and finely developed sense of pictorial space (figure 10.2).[6] With a career that spanned a forty-three-year period from the 1860s through the opening years of the twentieth century, Bennett lovingly, and in great detail, photographed what would become the region's most popular tourist destination. In doing so, he created an imaginative geography of picturesque refinement out of a space that, to most observers, at first seemed utterly ordinary. Part businessman, part inventor and technician, part artist and civic booster, Bennett embodied many of the tensions and contradictions that mark the intersection of rivers, photography, and tourism.

From Making Portraits to Viewing the River

H. H. Bennett's chosen form of graphic representation—the photographic view—was a peculiar class of photography, defined by neither subject, style, nor function but by the particular way that it integrated all three. Many of the still-dominant approaches to scenic or outdoor photography (as we see, for

Figure 10.1. "H. H. Bennett at Sugar Bowl." Unknown photographer. Wisconsin Historical Society, Image ID 8264.

example, in the work of Ansel Adams or Eliot Porter) were invented and refined during the three decades following the Civil War.[7] More than merely a pretty picture, the photographic view was a powerful American tradition, closely linked to larger forces of American capitalism and imperialism as the country stretched into new territories, organized and measured the land, colonized space, and transformed the landscape. Urbanization, industrialization, mechanization, and nationalization—these sweeping processes fundamentally transformed not only economic and social relations but cultural

Figure 10.2. "Sugar Bowl." H. H. Bennett. Wisconsin Historical Society, Image ID 7365.

perceptions as well, as Gilded Age Americans felt the shock of unparalleled changes. The new spaces generated in their wake required more than simple appropriation, as both wild river and urban space had to be ordered, interpreted, and situated within an American cultural hierarchy.[8]

View photography became the principal visual medium for this cultural work during the 1870s and 1880s. Although photographers had focused their cameras on American rivers before the Civil War and would continue to do so in the next century, the novelty and mysticism associated with this new representational process diminished over time, and it could no longer claim to be a "unique vehicle for American cultural expression."[9] Bennett came to view photography at just the right moment to reap the benefits of its ascendance as an expressive art form.

Just as pleasure travel expanded significantly in the late nineteenth century to include the emerging middle classes, so too did photography. The larger audiences that began to appreciate photography included many tourists.[10] Moreover, the technology of such mass-produced, low-cost images as the tintype, the carte de visite, and especially the stereograph permitted a far greater range of geographical subjects. With this democratization of sight and increasing mechanization came the contradictory—but mutually reinforcing —drive to produce images of American space that were more artistic and composed than previous ones had been.[11]

In 1865, upon returning to Wisconsin after the Civil War, Bennett purchased a local photographic business and began to work largely in portraiture.[12] Diary entries and letters to relatives from these earliest years indicate that the young veteran found portrait photography neither remunerative nor rewarding. He frequently wrote of the "victims" who posed for his camera and of his adopted town, Kilbourn City (now Wisconsin Dells), as a "*little, insignificant, dull, out of the way,* place," one that he doubted could support "business enough to give more than one person a decent living."[13] Only by supplementing his income with woodcutting, lathing work, construction for the railroad, and his Civil War pension did a "depressed" Bennett stave off economic destitution.[14]

The young photographer saw his fortunes change with his shift away from portraits and toward views of the nearby Wisconsin River. Although he never became wealthy, view photography along the Wisconsin provided Bennett a modest livelihood.[15] From the late 1860s through the first years of the twentieth century, he endlessly photographed and rephotographed sandstone formations and other natural features of the Dells that, in many cases, he himself named.[16] Like other view photographers, Bennett maintained a long-term contract with the railroad, for which he produced stereographs and mammoth prints for train stations. In 1876, for instance, the Chicago, Milwaukee, and St. Paul Railway distributed six thousand of Bennett's stereographs to libraries in the South to promote tourism, and in 1890 the Wisconsin Central Railroad commissioned Bennett to photograph Wisconsin resorts along its lines.[17] Mostly, however, he operated independently out of his small, tourist-oriented studio, where he sold stereo views and card photos of the Wisconsin River along with guidebooks and other souvenirs. The Bennett Studio, located conveniently near the railroad depot and riverboat docks, became an unofficial chamber of commerce where tourists went to obtain information about lodging and local attractions as well as to attend Bennett's lantern-slide lectures on river scenery and Native Americans.[18]

Bennett's busiest and most productive years coincided with the Wisconsin River Dells' rise to regional and national prominence, and his photographs contributed substantially to that ascension. During the 1870s the number of views that his studio could offer the tourist multiplied tenfold, from two hundred in 1872 to more than twenty-five hundred by the end of the decade.[19] As technical capabilities for photomechanical reproduction increased during the 1880s, he and his family were able to produce forty thousand prints per month.[20] By this time, Bennett's business had expanded well beyond drop-in visitors to his studio, and he was filling orders from places as far away as New York, Boston, and St. Louis.[21]

Reading the River in Photographs

Unlike many photographers of his day, H. H. Bennett received no formal education in the visual and literary arts. Whereas William Henry Jackson, for example, acquired his visual training from New England landscape painters and his intellectual notions of nature and landscape from Emerson's essays and Longfellow's poems, Bennett learned by creating the mass-produced stereographs that he sold in his studio and by observing the rural Wisconsin countryside.[22] This is not to suggest that Bennett's work remained uninfluenced by contemporary artistic conventions and economic pressures; to be sure, his photographs were strongly shaped by both. Rather, Bennett's photographs, more so even than Jackson's, were not just representational images but material manifestations of an emerging mass culture. Bennett frequently read the *Philadelphia Photographer* and the *Photographic Times*, the leading photographic journals of the day; his chief patron even contributed to the latter.[23] Moreover, the subjects he chose, like those of other view photographers, were influenced by buyer demand for the best-known geological features described in guidebooks and seen while touring along the river.[24] The photographic artifacts that emerged from H. H. Bennett's camera, then, bespoke a confluence of art and business, of photography and capitalism that appeared to transmit forthright information about the river. That his photographs were intended to satisfy his tourist clients and his own artistic sensibilities posed no contradiction for Bennett, nor did the fact that this most "modern" invention served to reinforce Victorian notions of gentility and refinement. For Bennett, such a cultural practice focused on remaking "America's hardest-working river" into a picturesque riverscape.[25]

In his skilled hands the Wisconsin River became "a fairy-story landscape, rugged and wild in half-scale, with enchanted miniature mountains and cool dark caves" (figure 10.3).[26] Stripped of any hint of social or environmental tension, Bennett's Dells assumed an air of tranquility and calm. The sculpted

Figure 10.3. "Out of Boat Cave." H. H. Bennett. Wisconsin Historical Society, Image ID 1970.

sandstone formations along the river became subjects for an outdoor portraiture, which, when purchased as a series, mimicked a boating trip up the Wisconsin River. The sublime beauty that had so impressed Bennett when he first encountered the river melted into a semi-wild playscape, "a sweet and not too dangerous place during the good months," John Szarkowski notes, "adventurous in aspect, but mapped and settled and free of wild Indians."[27]

Through the deep space of the stereo view, Bennett carefully traced the tourist's path along the river, from sight to sight, and recorded with fastidious

Figure 10.4. "Rowboat in Front of Eaton Grotto." H. H. Bennett. Wisconsin Historical Society, Image ID 8152.

care the picturesque sensibility.[28] Foregrounded elements—empty canoes resting on a sandy riverbank, tree branches outlining the edge of an image, a steamboat paralleling the river's rocky shore—seemed to leap out at the viewer, creating an illusion of depth and movement (figure 10.4). Most often Bennett achieved his aesthetic effects by peopling these riverscapes with well-dressed Victorian men, women, and children. Perched on overhanging rocks high above the river, picnicking on a sandbar, contemplating the grandeur of nature from an isolated point of land, or floating lazily downstream, their affluent presence conveyed the message that this landscape could be experienced safely and thus inspirationally (figure 10.5). While people do not seem to labor along Bennett's river, considerable ideological freight is carried by those recreating Victorians (figure 10.6). Local farmers, storekeepers, and merchants are nowhere to be seen. Tourists—people whose only relation to the river is to admire its beauty—claim this landscape as their own.

Figure 10.5. "Foot of High Rock." H. H. Bennett. Wisconsin Historical Society, Image ID 7894.

Replacing the hostility of the rapids with a frequently whimsical hospitality, Bennett created images of a welcoming natural environment that confirmed Victorian notions of gentility and refinement.[29] Bennett's Wisconsin River, though a place of retreat, was intended to edify and uplift. As one guidebook writer put it, *leisure* literally meant "to fix in the mind the beauties that are to be carried away for future reflection."[30] Even the scale of the riverscape seemed reduced to the size of a playground, providing the ideal setting for an escape from the city (figure 10.7). The genteel charm of this

Figure 10.6. "Berry's Landing." H. H. Bennett. Wisconsin Historical Society, Image ID 7343.

imaginary geography evoked notions of a place where nature could be encountered on the middle-class terms of control, safety, and inspiring recreation.[31]

A crucial aspect of Bennett's photographic work in the Dells was his naming of its specific features. Here his dual role as photographer and tourist developer is visible, as both enterprises benefited from the poetic names he chose for sandstone formations along a six-mile stretch of the Wisconsin River. As Bennett's son explained, "Many of his photographs necessitated his naming the places in the dells."[32] By calling attention to a rock's curious features and bestowing a colorful appellation on it, Bennett enhanced both the uniqueness of his photographic views as well as the river's tourism potential. Of the nearly ninety names he gave to features along the waterway, a few (such as Lone Rock, High Rock, or Notch Rock) described their objects literally, while most others were meant to evoke poetic associations: Lovers' Retreat, Angel Rock, Witches Gulch, Cave of the Dark Waters, Navy Yard, Echo Cove, Sugar Bowl, Fat Man's Misery, and Eagle Point (figure 10.8). Naming and viewing, Alan Trachtenberg has observed, complement each other. In Bennett's case, naming the river's features was not meant to incorporate the alien into a familiar system of knowledge; rather, he sought to enliven an otherwise static set of landmarks and to prepare the viewer for a more memorable tour of the river, whether by armchair or by steamboat.[33]

Bennett used the medium of photography to transform a haphazardly developed wilderness into the cultural myth of a pristine nature whose beauty and serenity could renew the human spirit.[34] This agricultural region of overworked wheat farms, poorly drained sand plains, hastily built towns, and quickly set rails showed the effects of labor unrest and recent Indian removals, but Bennett's work helped convert it into a warm-weather retreat for well-to-do Chicagoans.[35] This artful conversion was cloaked by photography's apparent scientific accuracy and unmediated truth. Although an earlier generation of photographers had sought to suppress the artificial character of the form and thus to provide a "realistic" window on nature, Bennett and his contemporaries focused on photography's conformity to the "rules of art and laws of composition."[36] It was precisely the artifice of the picture, its constructed quality, that, while appearing to be "natural," also suggested the ultimate human control over nature.

The figures in Bennett's photographs, as in so many Hudson River School paintings, gaze upon this landscape with romantic eyes; they invite the viewer to share in an inspiring experience of wilderness (see figure 10.2). But if American landscape painting most frequently idealized a middle, or pastoral landscape, Bennett's photographs recorded the passage into a postfrontier

Figure 10.7. "People on top of Stand Rock." H. H. Bennett. Wisconsin Historical Society, Image ID 7513.

world. Inspiration itself was mediated not by discovery but by recreation and the recognition that these beautiful vistas and natural phenomena could continually refresh the spirits of visitors from the cities. "As the frontier recedes," William Cronon suggests, "the wilderness ceases to be either an opportunity for progress or an occasion for terror. Instead, it becomes scenery."[37]

At the center of this new space was the transformation of *river* into *riverscape*, a transformation that was recorded within the prevailing discourse of photographic practices. Beginning in the 1880s, the word *scenery* began to occur with increasing frequency in important trade journals like *Anthony's* or the *Philadelphia Photographer*, eventually seeming to supplant the term *nature*. Riverscapes, for classic view photographers like Bennett, did not include images of labor or industry. Instead they represented America's waterways as playgrounds where scenic vistas and leisure activities were enjoyed by the burgeoning urban middle classes.[38]

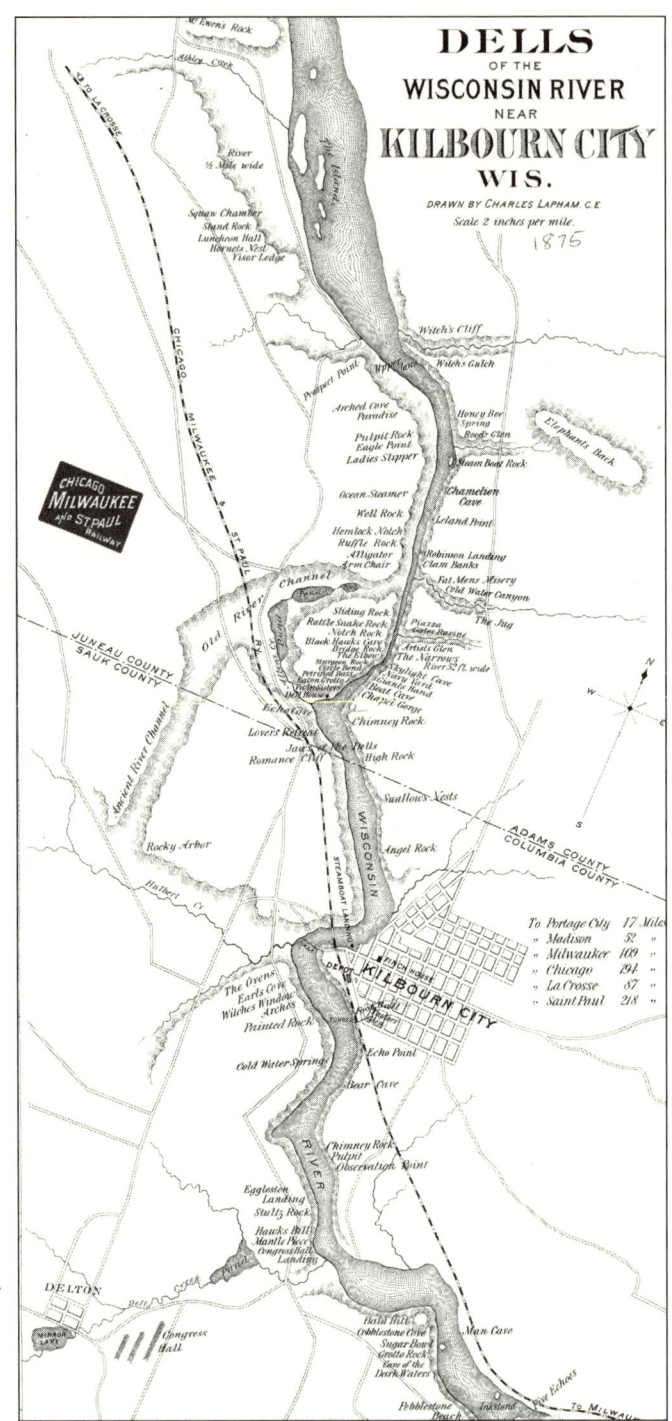

Figure 10.8. *Map of the Dells of the Wisconsin River near Kilbourn City, Wisconsin* (1891). Charles Lapham. Originally issued by Chicago, Milwaukee, and St. Paul Railway. Wisconsin Historical Society, Image ID 39790.

Labor and Play in a Scenic Riverscape

Bennett's postfrontier aesthetic could have evolved only by distortion of the contentious local history of the lumber industry and the ongoing dispossession and displacement of the native population. Although lumbermen still traveled through the Dells with their rafts of white pine and although the Winnebago, or Ho-Chunk, Indians continued to reside there, by the 1880s tourism had rendered both groups largely invisible in the area. Developers like Bennett could now represent these groups in ways that would further their own ends.[39] Bennett studiously photographed Native Americans and lumbermen, but rather than exposing these regional conflicts, Bennett's images suppressed them.[40]

This argument is illustrated in Bennett's series on the timber trade, which successfully converted an aqueous timber highway into a riverscape. Comprising forty stereographs taken mainly in October 1886 and entitled *The Camera's Story of Raftsman's Life on the Wisconsin*, the series depicts an industry in decline. Beginning in the 1820s and continuing for the next sixty years, the region's rivers served as highways on which raftsmen floated timber to markets in the south. But by the turn of the century, the railroad had rendered timber transport on the Wisconsin River obsolete, and Bennett wanted to make a visual record of this "vanishing" industry. He arranged with a Grand Rapids lumber company to join one of its logging crews on an eight-day trip from Kilbourn City to a point near the river's confluence with the Mississippi. Documenting the industry for posterity was not Bennett's primary aim; rather, he hoped to create views that he could then market to Milwaukee and Chicago lumber companies as well as to his expanding tourist clientele.[41]

Bennett's *Raftsman's* series presents a colorful and highly sentimental view of the industry. The violence and other social transgressions associated with rafting on the river remained beyond the margins of Bennett's camera. Mid-nineteenth-century newspapers brim with accounts of "mob-ocracy," of the "overpowering force of the destructionists," and of lumbermen "with small brains" who repeatedly destroyed the Kilbourn City dam, which they considered a hazard to rafting the Wisconsin River. Bennett's stereo views depicted none of these tensions but instead emphasized the virtues of manual labor and teamwork. His carefully posed scenes depict work, to be sure, but a wholesome work disconnected from the anarchistic behavior some raftsmen practiced and from the alienation of labor in an emerging urban-industrial order. It is work made to seem both manly and playful, a point that would not have been lost on nineteenth-century viewers (figure 10.9). This series was produced, after all, in the same year as Chicago's infamous Haymarket Riot, a turning point in the tumultuous labor relations that characterized the Gilded Age.[42]

The contrast between Chicago's new racially charged and volatile industrial order and Bennett's nostalgic images of the Dells emerged in bold relief

Figure 10.9. "Muscular Vigor in Repose" (*Raftsman's* series). H. H. Bennett. Wisconsin Historical Society, Image ID 7617.

Figure 10.10. "We Are Broke Up—Take Our Line" (*Raftsman's* series). H. H. Bennett. Wisconsin Historical Society, Image ID 7632.

through the *Raftsman's* series. It is important to note that Bennett conceived this stereo group as a distinct series. Purchased as a set of forty stereographs, the series was intended to be read as a narrative, with a beginning, middle, and end. The first few stereographs introduce the viewer to individual crew members, including the pilot of the logging fleet, and concludes with the fleet's safe arrival in Boscobel. The various scenes of the romantic drama unfold as Bennett illustrates aspects of a lifestyle then on the verge of extinction. Augmented with text descriptions such as "Handspiking Off a Sand Bar," or "Putting Down the Grouzers," or "We Are Broke Up—Take our Line" (figure 10.10), these visual narratives of physical labor reinforce the obsolescence of the entire enterprise. Every frontier story needs its benchmarks, and in the *Raftsmen* series Bennett created a polished vignette of the area's mythic origins. Indeed, by contrasting a view such as "Lumber Rafts in the Narrows" with "Apollo No. 1 Steamboat in Cold Water Canyon," one can readily gauge the temporal distance between a past devoted to resource extraction and a tourist-oriented present (figures 10.11 and 10.12). Both the mythic past and the playful present were crucial elements of the postfrontier aesthetic.

Figure 10.11. "Lumber Rafts in Narrows." H. H. Bennett. Wisconsin Historical Society, Image ID 6686.

"There is but little profit in the business"

Although H. H. Bennett followed well-established pictorial conventions, to read his photographs as commensurate with paintings fails to recognize the unique cultural functions performed by the professional view photographer, which were quite distinct from those of the landscape painter. The "discursive spaces" occupied by each differed significantly.[43] Exhibitions, for example, were rare for view photographers, whose work was driven overwhelmingly by market concerns. Bennett's riverscapes, produced for a large and diverse audience, whether in stereoscopic or mammoth plate format, were thus highly responsive to changing demands and tastes. Equally important, changes in the marketing and distribution process, along with techno-

Figure 10.12. "Apollo No. 1 Steamboat in Cold Water Canyon." H. H. Bennett. Modern print from original 8 x 10 inch glass negative (c. 1900). Wisconsin Historical Society, Image ID 7883.

logical developments, entwined with the photographs themselves and with the practices that surrounded their works' creation and consumption.[44]

In a letter of 1899, a disheartened Bennett wrote, "My advice to the young would be not to learn photography as a means of gaining a livelihood." He further lamented that "the hard times together with so many people buying the cheap snap shot cameras has brought about this condition of things. I cannot say how long a time is required to become a good artist [as] some pick it up in a comparatively short time, while others never get to know it at all. It's a work that is continually changing." Especially for photographers in "resort regions" along America's rivers, Bennett concluded, "competition is so keen and prices for pictures so low that there is but very little profit in the business."[45] In other words, the view photographer's career rested on a precarious balance of economic, aesthetic, and cultural factors. Although Bennett had worked out of his rural Wisconsin studio successfully for thirty years and had earned national renown for his stereographic views of river scenery, by the turn of the century the "circuit of culture" linking photographers and viewers had altered dramatically.[46] The blending of artistic composition with techniques of mass production and distribution that had made him so successful after the Civil War was not enough to help Bennett meet the many new technical and commercial challenges that had arisen by 1900. The so-called amateur photography craze was just one obstacle. As the traditional distribution channels for his views began to disappear, demand continued to rise for photographs of the now-numerous tourist excursions on the river. But soon this appetite, too, was exhausted as big capital came to control the boating operations (thus cutting Bennett out of the excursion trade) and the form of tourism itself gradually shifted from gazing at natural scenery to pursuing amusements or engaging in other recreational diversions.[47]

Furthermore, within five years halftone and color-tinted postcards replaced virtually all other photographic media at the Bennett studio, including stereographs.[48] The more cost-effective reproduction technologies used for postcards discouraged mass production of original river views as huge publishing conglomerates squeezed smaller entrepreneurs out of the marketplace or relegated the photographer to the role of employee.[49] As a result, photographs began to be regarded less as art objects and more as enhancements of written text (figure 10.13). Tinting of photographs grew in popularity, image sizes were greatly reduced, and the Phostint process recently imported from Switzerland produced an impressionistic, continuous-tone appearance.[50] This modernized system of manufacturing visual artifacts undercut

Figure 10.13. *Dells of the Wisconsin River at Kilbourn City.* Postcard printed by E. C. Kropp with photograph by H. H. Bennett (c. 1905). Author's collection.

the Gilded Age notion of the view photograph's window on a transcendent—if artfully created—reality.

These powerful forces of technological and economic change disrupted the tenuous connection between commerce and art that had made possible the impressive successes of river photographers like H. H. Bennett. Likewise, the alliance between view photography and nature as a postfrontier tourist space dissolved. Out of this disintegration, new alliances formed between photography and other strands of American culture and between tourism and visual media. As the leading edge of view photography moved toward a separation of photography's commercial viability from its aesthetic imperatives—exemplified most readily by such modernists as Alfred Stieglitz and Gertrude Käsebier—tourism in places like the Wisconsin River Dells leaned toward mass-culture forms of entertainment and away from the contemplation of picturesque views along the river that had so moved Gilded Age tourists. By the time of his death in 1908, Bennett's focus had shifted away from original photography of the river to supplying the ever-greater demand by tourists for Indian souvenirs and postcards.[51]

Tourists still travel through the Dells, and they continue to visit the H. H. Bennett photography studio, though both spaces have been radically changed. Wisconsin's first great hydroelectric project—the damming of the river at the Dells in 1909—was not a financial success, but it did succeed in submerging many of the geological features made famous by Bennett's photographic views.[52] Boat companies today plead with tourists to "see first things first," and, while some do visit the river, most visitors prefer the artificial waterway of one of the area's more than two dozen waterparks, including the gargantuan Noah's Ark Waterpark. The Bennett studio has fared somewhat better than the river itself, though its transformation has been no less complete. A working photographic studio until 1999, when it was acquired by the Wisconsin Historical Society, the studio is now a flourishing state historic museum that offers an opportunity to learn about the river during the nineteenth century. The river's earlier life, as this chapter has sought to demonstrate, is well illustrated—indeed, strongly influenced—by the artful photographs of H. H. Bennett.

These photographic views reflected and helped to promote a nineteenth-century tourist aesthetic based on the belief that rivers afforded opportunities for recreation and inspiration, that waterways were not merely conduits for timber but also resorts for vacationing Victorians. The American Midwest had, after all, already witnessed a half century of intensive use and alteration during which the surrounding countryside had become well integrated into a vast urban-focused network of railroads, resources, and markets. The frontiers of white settlement and of timbering had long since passed, turning a zone of intercultural conflict into an increasingly modern, accessible region of modest farms and even more modest towns and cities. This new kind of space—as foreign to the Gilded Age experience as the exploding metropolis itself—combined the pseudo-wild riverscape with new social hierarchies and consumer-driven enterprises.

While H. H. Bennett himself may have credited God or natural forces for the physical construction of the river that he so tirelessly photographed, his own work in a very real sense created the Dells. Visual images, no less than language, can play an active role in the production of meaning. Photographs direct our attention to things formerly overlooked—and thus invisible and nonexistent—as they simultaneously organize insignificant entities into significant composite wholes.[53] Bennett's accomplishment lay precisely in his ability to craft such significant composite wholes out of the disparate geological formations of the river and translate them into a coherent image of genteel recreation. A working river with a long history of timber-rafting and sawmilling activity, as well as of anarchy and violence, was transformed into a picturesque landscape—a riverscape—largely through the medium of the view photograph.

Notes

Chapter 1: Rivers in History and Historiography

1. Lewis Mumford, *Technics and Civilization* (New York: Harcourt, Brace, Jovanovich, 1934), 60–61.

2. Mark Cioc, *The Rhine: An Eco-Biography, 1815–2000* (Seattle: University of Washington Press, 2002), 207; Simon Schama, *Landscape and Memory* (New York: Vintage Books, 1996), 355ff.

3. For analyses of these valorizations, see Lucien Febvre and Albert Demangeon, *Le Rhin* (Paris: A. Colin, 1935); and, more recently, Jonathan Schneer, *Thames: A Biography* (New Haven: Yale University Press, 2005). On the Netherlands, see the following essays in the July 2002 issue of *Technology and Culture* (vol. 43, no. 3): William H. TeBrake, "Taming the Waterwolf: Hydraulic Engineering and Water Management in the Netherlands during the Middle Ages," 475–99; Petra J. E. M. van Dam, "Ecological Challenges, Technological Innovations: The Modernization of Sluice-Building in Holland, 1300–1600," 500–520; Arne Kaijser, "System-Building from Below: Institutional Change in Dutch Water-Control Systems," 521–48; Harry Lintsen, "Two Centuries of Central Water Management in the Netherlands," 549–68; and Wiebe E. Bijker, "The Oosterschelde Storm-Surge Barrier: A Test Case for Dutch Water Technology, Management, and Politics," 569–84.

4. William Cronon, "Time and the River Flowing," foreword to *The Rhine: An Eco-Biography, 1815–2000*, by Mark Cioc (Seattle: University of Washington Press, 2002), ix–xii, esp. xi.

5. See, for example Peter Calow and Geoffrey E. Petts, eds., *The Rivers Handbook: Hydrological and Ecological Principles*, 2 vols. (Oxford: Blackwell Scientific, 1992–94).

6. T. T. Veblen and D. C. Lorenz, *The Colorado Front Range: A Century of Ecological Change* (Salt Lake City: University of Utah Press, 1991); Ellen E. Wohl, *Virtual Rivers: Lessons from the Mountain Rivers of the Colorado Front Range* (New Haven: Yale University Press, 2001), 9–15, 182–83. See also Christian Pfister and Daniel Brändli, "Rodungen im Gebirge, Überschwemmungen im Vorland: Ein Deutungsmuster macht Karriere," in *Natur-Bilder: Wahrnehmungen von Natur und Umwelt in der Geschichte*, ed. Rolf Peter Sieferle and Helga Breuninger (Frankfurt: Campus, 1999), 297–324.

7. Robert C. Post, *Technology, Transport, and Travel in American History* (Washington, DC: American Historical Association/Society for the History of Technology, 2003), 28. On the best-known canal in the United States, see Carol Sheriff, *The Artificial River: The Erie Canal and the Paradox of Progress, 1817–1862* (New York: Hill and Wang, 1997).

8. Cioc, *The Rhine*, 38–39; and David Blachbourn, *The Conquest of Nature: Water, Landscape, and the Making of Modern Germany* (New York: Norton, 2006), 103–4.

9. For recent studies of the Corps's activities, see Martin Reuss, *Designing the Bayous: The Control of Water in the Atchafalaya Basin, 1800–1995* (College Station: Texas A&M University Press, 2004); and Karen M. O'Neill, *Rivers by Design: State Power and the Origins of U.S. Flood Control* (Durham, NC: Duke University Press, 2006).

10. Post, *Technology, Transport, and Travel*, 41.

11. Peirce F. Lewis, *New Orleans: The Making of an Urban Landscape* (Cambridge, MA: Ballinger, 1976); Philip V. Scarpino, *Great River: An Environmental History of the Upper Mississippi, 1890–1950* (Columbia: University of Missouri Press, 1985); John O. Anfinson, *The River We Have Wrought: A History of the Upper Mississippi* (Minneapolis: University of Minnesota Press, 2003); Ari Kelman, *A River and Its City: The Nature of Landscape in New Orleans* (Berkeley: University of California Press, 2003); Craig E. Colten, *An Unnatural Metropolis: Wresting New Orleans from Nature* (Baton Rouge: Louisiana State University Press, 2005).

12. For useful post-Katrina assessments, see the following short essays in the January 2006 issue of *Technology and Culture* (vol. 47, no. 1): Todd A. Shallat, "Holding Louisiana," 102; Carolyn Kolb, "Crescent City, Post-Apocalypse," 108; and Barbara L. Allen, "Cradle of a Revolution? The Industrial Transformation of Louisiana's Lower Mississippi River," 112.

13. See Donald Worster, *Rivers of Empire: Water, Aridity, and the Growth of the American West* (New York: Oxford University Press, 1985); Donald J. Pisani, *To Reclaim a Divided West: Water, Law, and Public Policy, 1848–1902* (Albuquerque: University of New Mexico Press, 1992); Donald J. Pisani, *Water and American Government: The Reclamation Bureau, National Water Policy, and the West, 1902–1935* (Berkeley: University of California Press, 2002); and David E. Nye, *American Technological Sublime* (Cambridge, MA: MIT Press, 1994), 137–42.

14. World Commission on Dams, *Dams and Development: A New Framework for Decision-Making: The Report of the World Commission on Dams* (Sterling, VA: Earthscan, 2000); Paul R. Josephson, *Resources under Regimes: Technology, Environment, and the State* (Cambridge, MA: Harvard University Press, 2004).

15. Loren R. Graham, *What Have We Learned about Science and Technology from the Russian Experience?* (Stanford, CA: Stanford University Press, 1998), 120. As this book goes to press, environmentalists are protesting another controversial large-scale dam, the Sardar Sarovar, which is part of the Narmada Dam project in India.

16. John Wesley Powell, *Report on the Lands of the Arid Regions of the United States* (Washington, DC: Government Printing Office, 1878). See also John Wesley Powell, *Report of the Survey of the Colorado of the West: Letter from the Secretary of the Smithsonian Institution, Transmitting a Report of the Survey of the Colorado of the West and, Its Tributaries* (Washington, DC: Government Printing Office, 1873).

17. See Karl A. Wittfogel and Fêng Chia-shêng, *History of Chinese Society: Liao, 907–1125* (Philadelphia: American Philosophical Society, 1949); Gary L. Ulmen, *The Science of Society: Toward an Understanding of the Life and Work of Karl August Wittfogel* (The Hague and New York: Mouton, 1978); and Worster, *Rivers of Empire*, 22–29.

18. In particular, see Donald J. Pisani's *To Reclaim a Divided West* and *Water and American Government.* In the context of the Netherlands, expansive control of water appears to have led not to centralized governments but to local, cooperative control for several centuries.

19. See Bill McKibben, *The End of Nature* (New York: Random House, 1989); Patrick McCully, *Silenced Rivers: The Ecology and Politics of Large Dams* (Atlantic Highlands, NJ: Zed Books, 1996); and Eva Jakobsson, "How Do Historians of Technology and Environmental Historians Conceive the Harnessed River?" lecture delivered at "Rivers in History: Designing and Conceiving Waterways in Europe and North America" conference, December 4–7, 2003, German Historical Institute, Washington, DC.

20. In addition to the Jakobsson lecture cited above, see also Jakobsson, *Industrialisering av älvar: Studier kring svensk vattenkraftutbyggnad 1900–1918* (Göteborg: Göteborgs Universitet, 1996); and Jakobsson, "Industrialised Rivers: The Development of Swedish Hydro-Power," in *Nordic Energy Systems: Historical Perspectives and Current Issues,* ed. Arne Kaijser and Marika Hedin (Canton, MA: Science History Publications/USA, 1995).

21. Philip L. Fradkin, *A River No More: The Colorado River and the West,* 2nd ed. (Berkeley: University of California Press, 1996), xxii.

22. Wohl, *Virtual Rivers,* x. Richard White also uses the term *virtual river* to refer to the Columbia River in his book *The Organic Machine* (New York: Hill and Wang, 1995), 108.

23. White, *The Organic Machine,* 60.

24. Ibid, 109.

25. Ibid.

26. Cioc, *The Rhine,* 5.

27. Ibid, 6.

28. Marc Reisner, *Cadillac Desert: The American West and Its Disappearing Water* (New York: Penguin Books, 1993); Pisani, *To Reclaim a Divided West;* Norris Hundley Jr., *Water and the West: The Colorado River Compact and the Politics of Water in the American West* (Berkeley: University of California Press, 1975); Norris Hundley Jr., *The Great Thirst: Californians and Water—A History* (Berkeley: University of California Press, 2001); Charles F. Wilkinson, *Crossing the Next Meridian: Land, Water, and the Future of the West* (Washington, DC: Island Press, 1992).

29. For a German study focusing on pollution, see Jürgen Büschenfeld, *Flüsse und Kloaken: Umweltfragen im Zeitalter der Industrialisierung (1870–1918)* (Stuttgart: Klett-Cotta, 1997). On riverine pollution in Britain, see Bill Luckin, *Pollution and Control: A Social History of the Thames in the Nineteenth Century* (Bristol and Boston: Adam Hilger, 1986); and Christopher S. Hamlin, *A Science of Impurity: Water Analysis in Nineteenth-Century Britain* (Berkeley: University of California Press, 1990).

30. Since the submission of the present collection, a three-volume work emanating from conferences organized by the International Water History Association has been published under the title *A History of Water* (London: I. B. Tauris, 2006). *Water Control and River Biographies,* vol. 1, ed. Terje Tvedt and Eva Jakobsson; *The Political Economy of Water,* vol. 2, ed. Richard Coopey and Terje Tvedt; *The World of Water,* vol. 3, ed. Terje Tvedt and Terje Oestigaard.

31. Marcus Hall, *Earth Repair: A Transatlantic History of Environmental Restoration* (Charlottesville: University of Virginia Press, 2005); Daniel P. Loucks, ed., *Restoration of Degraded Rivers: Challenges, Issues, and Experiences* (Dordrecht: Kluwer, 1998).

32. For forays into the history of Chinese rivers, see Robert Marks, *Tigers, Rice, Silk, and Silt: Environment and Economy in Late Imperial South China* (Cambridge: Cambridge University Press, 1998); Lyman P. Van Slyke, *Yangtze: Nature, History, and the River* (Reading, MA: Addison-Wesley, 1988); Mark Elvin and Ts'ui-jung Liu, eds., *Sediments of Time: Environment and Society in Chinese History* (Cambridge: Cambridge University Press, 1998), chaps. 2, 5, 9, 10, 15, 18, 20. The editors are indebted to Andrea Goldman (UCLA, formerly of the University of Maryland) for this reference. On cultural meanings of the Amazon, see Candace Slater, *Entangled Edens: Visions of the Amazon* (Berkeley: University of California Press, 2002). On the Nile, see Karl W. Butzer, *Early Hydraulic Civilization in Egypt: A Study in Cultural Ecology* (Chicago: University of Chicago Press, 1976); and Terje Tvedt, *The River Nile in the Age of the British: Political Ecology and the Quest for Economic Power* (London: I. B. Tauris, 2004). Meredith McKittrick of Georgetown University is currently completing a historical study of conflicts over shared water resources in the six major river systems of southern Africa.

Chapter 2: "Time is a violent torrent"

1. T. S. Eliot, "The Dry Salvages," in *The Four Quartets* (London: Faber and Faber, 1944), ll. 1–2.

2. Heraclitus, in Plato, *Cratylus*, 402a; Marcus Aurelius, *Meditations*, Bk. 4, §43.

3. On dikes and revolution, see Klemens von Metternich, cited in James J. Sheehan, *German History, 1770–1866* (Oxford: Clarendon Press, 1989), 604 (on 1830); Karl Biedermann, *Mein Leben und ein Stück Zeitgeschichte*, 2 vols. (Breslau: Schottlaender, 1886), 1:278–79 (on 1848); Josef Maria von Radowitz to Marie von Radowitz, Mar. 3, 1848, in *Nachgelassene Briefe und Aufzeichnungen zur Geschichte der Jahre 1848–1853*, ed. Walter Möring (Stuttgart: Deutsche Verlags-Anstalt, 1922), 9; and on the Slav "flood," see Wolfgang Wippermann, *Der "Deutsche Drang nach Osten": Ideologie und Wirklichkeit eines politischen Schlagwortes* (Darmstadt: Wissenschaftliche Buchgesellschaft, 1981), 98–99.

4. Niccolò Machiavelli, *The Prince*, trans. Harvey C. Mansfield Jr. (Chicago: University of Chicago Press, 1985), 98.

5. Roger D. Masters, *Fortune Is a River: Leonardo da Vinci and Niccolò Machiavelli's Magnificent Dream to Change the Course of Florentine History* (New York: Free Press, 1998).

6. Charles Tilly, *Big Structures, Large Processes, Huge Comparisons* (New York: Sage, 1984).

7. Clifford Geertz, *Local Knowledge: Further Essays in Interpretive Anthropology* (New York: Basic Books, 1983).

8. John McNeill, *Something New under the Sun: An Environmental History of the Twentieth Century* (New York: Norton, 2000), 166–73; Raimund Rödel, *Die Auswirkungen des historischen Talsperrenbaus auf die Zuflussverhältnisse der Ostsee* (Greifswald: Ernst-Moritz-Arndt-Universität Greifswald, 2001).

9. Johann Christoph Bekmann, quoted in Theodor Fontane, *Wanderungen durch die Mark Brandenburg*, 3 vols. (Munich: Hanser Verlag, 1991), 1:566.

10. Ibid., 1:550; Walter Christiani, *Das Oderbruch: Historische Skizze*, 3rd ed., rev. (Freienwalde: Pilger, 1901), 13. See also Johann Bernoulli, *Reisen durch Brandenburg, Pommern, Preussen . . . in den Jahren 1777 und 1778*, 6 vols. (Leipzig: Fritsch, 1779–80), 1:39.

11. Max Beheim-Schwarzbach, *Hohenzollernsche Colonisationen: Ein Beitrag zu der Geschichte des preussischen Staates und der Colonisation des östlichen Deutschlands* (Leipzig: Duncker & Humblot, 1874), 266.

12. August Gottlob Meissner, *Leben Franz Balthasar Schönberg von Brenkenhof* (Leipzig: Breitkopf, 1782), 80–81; Erich Neuhaus, *Die Fridericianische Colonisation im Netze- und Warthebruch* (Landsberg: Schaeffer, 1905), 8–9; Clarence J. Glacken, *Traces on the Rhodian Shore: Nature and Culture in Western Thought from Ancient Times to the End of the Eighteenth Century* (Berkeley: University of California Press, 1967), 604–6, 659–65, 680–81, 588–89, 702–3; Yi-Fu Tuan, *Passing Strange and Wonderful: Aesthetics, Nature, and Culture* (Washington, DC: Island Press, 1993), 61–77.

13. "Eulogium of Euler," in *Letters of Euler . . . Addressed to a German Princess*, trans. Henry Hunter, 2 vols. (London: Murray and Highley, 1802), 1:lxv; Karl Karmarsch, *Geschichte der Technologie seit der Mitte des 18. Jahrhunderts* (Munich: Oldenburg, 1872), 13–14; Günther Garbrecht, *Wasser: Vorrat, Bedarf und Nutzung in Geschichte und Gegenwart* ([Munich]: Deutsches Museum; Reinbek bei Hamburg: Rowohlt, 1985), 178–81.

14. Haerlem and Petri to Colonel von Retzow, July 18, 1753, quoted in Albert Detto, "Die Besiedlung des Oderbruches durch Friedrich den Großen," *Forschungen zur Brandenburgischen und Preussischen Geschichte* 16.1 (1903): 163–205, esp. 177.

15. Fontane, *Wanderungen*, 1:547; Reinhold Koser, *Geschichte Friedrich des Großen*, 3 vols. (Darmstadt: Wissenschaftliche Buchgesellschaft, 1974), 3:97.

16. Heuer's verse is on the unnumbered endpages of Christiani, *Oderbruch*.

17. Ibid., "Vorwort"; Ernst Breitkreutz, *Das Oderbruch im Wandel der Zeit: Ein kulturhistorisches Bild* (Remscheid: Privately printed, 1911), iii.

18. James C. Scott, *Seeing Like a State: How Certain Schemes to Improve the Human Condition Have Failed* (New Haven: Yale University Press, 1998).

19. Otto Kaplick, *Das Warthebruch: Eine deutsche Kulturlandschaft im Osten* (Würzburg: Holzner-Verlag, 1956), 23–25.

20. Breitkreutz, *Oderbruch*, 117.

21. Werner Michalsky, *Zur Geschichte des Oderbruchs: Die Besiedlung* (Seelow: Rat d. Kreises Seelow, Abt. Kultur, 1983), 12.

22. Christof Dipper, *Deutsche Geschichte 1648–1789* (Frankfurt: Suhrkamp, 1991), 10–18; Brian Fagan, *The Little Ice Age: How Climate Made History, 1300–1850* (New York: Basic Books, 2000).

23. Heinz Musall, *Die Entwicklung der Kulturlandschaft der Rheinniederung zwischen Karlsruhe und Speyer vom Ende des 16. bis zum Ende des 19. Jahrhunderts* (Heidelberg: Geographischen Instituts der Universität Heidelberg, 1969), 44, 53–57, 67–69, 169–71; Horst Johannes Tümmers, *Der Rhein: Ein europäischer Fluss und seine Geschichte* (Munich: C. H. Beck, 1994), 139–40.

24. Quoted in Hans Georg Zier, "Johann Gottfried Tulla: Ein Lebensbild," *Badische Heimat* 50 (1970): 379–449, esp. 399.

25. Tulla's *Denkschrift* of Mar. 1, 1812, quoted in Tümmers, *Der Rhein*, 145.

26. Johannes Gut, "Die badisch-französische sowie die badisch-bayerische Staatsgrenze und die Rheinkorrektion," *Zeitschrift für die Geschichte des Oberrheins* 142 (1994): 215–32; Christoph Bernhardt, "The Correction of the Upper Rhine in the Nineteenth Century: Modernizing Society and State by Large-Scale Water Engineering," in *Water, Culture, and Politics in Germany and the American West*, ed. Susan C. Anderson and Bruce H. Tabb (New York: Peter Lang, 2001), 183–202; Mark Cioc, *The Rhine: An Eco-Biography, 1815–2000* (Seattle: University of Washington Press, 2002), 49–50.

27. Quoted in Peter Sahlins, "Natural Frontiers Revisited: France's Boundaries since the Seventeenth Century," *American Historical Review* 95.5 (1990): 1423–51, esp. 1442.

28. Quoted in Zier, "Johann Gottfried Tulla," 431, 440.

29. Götz Kuhn, *Die Fischerei am Oberrhein: Geschichtliche Entwicklung und gegenwärtiger Stand* (Stuttgart: E. Ulmer, 1976); Anton Lelek and Günter Buhse, *Fische des Rheins: Früher und heute* (Berlin: Springer, 1992); Thomas Tittizer and Falk Krebs, eds., *Ökosystemforschung: Der Rhein und seine Auen—eine Bilanz* (Berlin: Springer, 1996); Cioc, *The Rhine*, 150–67.

30. See the findings of Ragnar Kinzelbach, "Veränderungen der Fauna im Oberrhein," and Georg Philippi, "Änderung der Flora und Vegetation am Oberrhein," both in *Natur und Landschaft am Oberrhein: Versuch einer Bilanz*, ed. Norbert Hailer (Speyer: Verlag der Pfälzischen Gesellschaft zur Förderung der Wissenschaften, 1982), 66–83 and 87–103, respectively. Tittizer and Krebs devote more than a hundred pages to the effects of pollution alone in their *Ökosystemforschung*.

31. Herbert Schwarzmann, "War die Tulla'sche Oberrheinkorrektion eine Fehlleistung im Hinblick auf ihre Auswirkungen?" *Die Wasserwirtschaft* 54 (1960): 279–87; Fritz Schulte-Mäter, *Beiträge über die geographischen Auswirkungen der Korrektion des Oberrheins* (Leipzig: Moltzen, 1938), 19–20, 27–38; Bernhardt, "Correction of the Upper Rhine," 192–99; Cioc, *The Rhine*, 33–36, 65, 69–75.

32. Quoted in Wolfgang Diehl, "Poesie und Dichtung der Rheinebene," in *Der Rhein und die Pfälzische Rheinebene*, ed. Michael Geiger, Günter Preuss, and Karl-Heinz Rothenberger (Landau in der Pfalz: Verlag Pfälzische Landeskunde, 1991), 378–93, esp. 384.

33. Quoted in Hans-Rüdiger Fluck, "Die Fischerei im Hanauerland," *Badische Heimat* 50.4 (1970): 466–89, esp. 484.

34. Michael J. Quin, *Steam Voyages on the Seine, the Moselle, & the Rhine*, 2 vols. (London: H. Colburn, 1843), 2:116. Compare Lucy Hill, *Rhine Roamings* (Boston: Lee and Shepard, 1880), 52. In 1840 Friedrich Engels satirized this guidebook-defined Rhine in an essay titled "Siegfrieds Heimat," reprinted in *Deutsche Landschaften*, ed. Helmut J. Schneider (Frankfurt: Insel Verlag, 1981).

35. Carl Borchardt, *Die Remscheider Stauweiheranlage sowie Beschreibung von 450 Stauweiheranlagen* (Munich: Oldenbourg, 1897); Axel Föhl and Manfred Hamm, *Die Industriegeschichte des Wassers* (Düsseldorf: VDI Verlag, 1985), 128.

36. Richard Hennig, *Buch Berühmter Ingenieure: Große Männer der Technik, ihr Lebensgang und ihr Lebenswerk; für die reifere Jugend und für Erwachsene geschildert* (Leipzig: Spamer, 1911), 104–21.

37. O. Bechstein, "Vom Ruhrtalsperrenverein," *Prometheus* 28 (Oct. 1916): 135–39, esp. 138; L. Ernst, "Die Riesentalsperre im Urftal [sic]," *Die Umschau* (1904): 667–68; Hermann Schönhoff, "Die Möhnetalsperre bei Soest," *Die Gartenlaube* (1912): 684–86.

38. Hans Dominik, *Im Wunderland der Technik* (Berlin: R. Bong, 1922); see also Bernhard Rieger, "'Modern Wonders': Technological Innovation and Public Ambivalence in Britain and Germany between the 1890s and 1933," *History Workshop Journal* 55 (2003): 154–78. David E. Nye, in *American Technological Sublime* (Cambridge, MA: MIT Press, 1994), argues that the European capacity for wonder in the face of the sublime remained stuck in the age of Kant and Burke. According to Nye, Europeans never made the leap that allowed them to celebrate the "technological sublime" as Americans did.

39. Kurt Soergel, "Die Bedeutung der Talsperren in Deutschland für die Landwirtschaft" (Ph.D. diss., University of Leipzig, 1929).

40. P. Ziegler, "Ueber die Notwendigkeit der Einbeziehung von Thalsperren in die Wasserwirtschaft," *Zeitschrift für Gewässerkunde* (hereafter cited as *ZfG*) (1901): 49–58, esp. 52.

41. Andreas Kunz, "Binnenschiffahrt," in *Technik und Wirtschaft,* ed. Ulrich Wengenroth (Düsseldorf: VDI Verlag, 1993), 385–87.

42. A complete transcription of Intze's address at the Marklissa dam appears in "Ueber Talsperren," *ZfG* 4 (1902): 252–54, esp. 253. See also Otto Intze, *Bericht über die Wasserverhältnisse der Gebirgsflüsse Schlesiens im Bober- und Queissgebiete sowie im Gebiete der Glatzer Neisse und deren Verbesserung zur Ausnutzung der Wasserkräfte sowie zur Verminderung der Hochwasserschäden durch Anlage von Sammelbecken* (Berlin: C. Heymann, 1899).

43. H. Christian Nussbaum, "Die Wassergewinnung durch Talsperren," *Zeitschrift für die Gesamte Wasserwirtschaft* (1907): 67–70. See also Paul Ziegler, *Der Talsperrenbau,* 2nd ed., rev. (Berlin: Wilhelm Ernst, 1911), 4–5; W. Berdrow, "Staudämme und Thalsperren," *Die Umschau* (1898): 255–59, esp. 256.

44. Soergel, "Bedeutung der Talsperren," 39.

45. The multiple is closer to eight hundred times when it comes to signature postwar dams like the Kariba, Volta, and High Aswan dams. For German high dams on the ICOLD register (311 out of more than 40,000), see Peter Franke, ed., *Dams in Germany* (Essen: Verlag Glückauf, 2001), 466–95; and Rolf Meurer, *Wasserbau und Wasserwirtschaft in Deutschland* (Berlin: Parey, 2000), 186.

46. See, for example, Rainer Blum, "Seismische Überwachung der Schlegeis-Talsperre und die Ursachen induzierte Seismizität" (Ph.D. diss., University of Karlsruhe, 1975); and David Alexander, *Natural Disasters* (New York: Chapman and Hall, 1993), 56–57, 359.

47. August Thienemann, "Hydrobiologische und fischereiliche Untersuchungen an den westfälischen Talsperren," *Landwirtschaftliche Jahrbücher* 41 (1911): 535–716; Thomas Kluge and Engelbert Schramm, *Wassernöte: Umwelt- und Sozialgeschichte des Trinkwassers* (Aachen: Alano Verlag, 1986), 169–72.

48. Richard White, "The Natures of Nature Writing," *Raritan* 22.2 (2002): 145–61, esp. 154–55.

49. Fischer-Reinau, "Die wirtschaftliche Ausnützung der Wasserkräfte," *Bayerisches Industrie- und Gewerbeblatt* (1908): 71–77, esp. 106. Compare Emil Abshoff, "Einiges über Talsperren," *Zeitschrift für Binnenschiffahrt* (1905): 91–92; Ernst Mattern, *Die Ausnutzung der Wasserkräfte: Technische und wirtschaftliche Grundlagen; neuere Bestrebungen der Kulturländer* (Leipzig: Engelmann, 1921), 1001–2.

50. Karl Kollbach, "Die Urft-Talsperre," *Über Land und Meer* 92 (1913–14): 694–95; W. Abercron, "Talsperren in der Landschaft: Nach Beobachtungen aus der Vogelschau," *Volk und Welt* (June 1938): 33–39.

51. Paul Schultze-Naumburg, "Kraftanlagen und Talsperren," *Der Kunstwart* 19 (1906): 130; J. Weber, "Die Wupper-Talsperren," *Bergische Heimat* 4 (Aug. 1930): 313–23, esp. 316.

52. August Trinius, quoted in William H. Rollins, *A Greener Vision of Home: Cultural Politics and Environmental Reform in the German Heimatschutz Movement, 1904–1918* (Ann Arbor: University of Michigan Press, 1997), 246.

53. Donald Worster, *The Wealth of Nature: Environmental History and the Ecological Imagination* (New York: Oxford University Press, 1993), 123–34.

54. Richard White, *The Organic Machine: The Remaking of the Columbia River* (New York: Hill and Wang, 1995), 112.

Chapter 3: From Parisian River to National Waterway

1. A comprehensive description of sources can be found in Isabelle Backouche, *La trace du fleuve: La Seine et Paris (1750–1850)* (Paris: Éditions de l'EHESS, 2000).

2. The bridges that crisscrossed Île de la Cité acted as powerful barriers, creating two types of navigation and two types of commercial exchange within Paris. Upstream, small light boats floated down the Seine to supply the city with commodities such as coal, wood, and grains, while downstream, oceangoing ships loaded with imported luxury goods were towed upriver to Paris.

3. For more information on the laundry boats, see the series of leases signed between the laundry boat managers and the landlords to whom the city had entrusted management of waterfront areas, preserved in the Archives Nationales de Paris (hereafter ANP), Q^{1}1118 and Q^{1}1119. On the mills, see "Ordonnance du prévôt des marchands pour le placement des bateaux à moulin," dated Dec. 28, 1787, ANP, H 1958.

4. "Lettres patentes du roi concernant l'établissement à former dans l'Isle des Cignes, pour le netoyement, cuisson et préparation des abattis, de bœufs, vaches et moutons," dated Jan. 7, 1763, ANP, AD XVI 10.

5. This calculation is derived from statistics on the raising of taxes and grants from the years 1787, 1788, and 1789, housed at the Archives de Paris, Section départementale, Domaines, carton 42. See also "Etat nominatif des receveurs établis par les fermiers généraux sur les ports et aux barrières de la ville de Paris . . . , n.d.," ANP, H 2175. It is difficult to estimate with any accuracy eighteenth-century water usage, but the subject of the city's water supply provoked many debates between those advocating a greater use of

the Seine and those in favor of other solutions, such as rerouting effluent rivers or building aqueducts.

6. "Ordonnance portant confirmation des privilèges, ordonnances et réglements sur la police de l'Hôtel de ville de Paris, et réglement sur la juridiction des prévôt des marchands et échevins," dated Dec. 1672, ANP, H 1914.

7. The first port, named Port de la Grève, designated one pier for wood, one for wheat, one for wine, and one for hay. By the eighteenth century, Port de la Grève had been superseded by ports upstream—such as Port Saint-Paul, which specialized in coal, wood, iron, grains, wine, chalk, and slaked lime; Port de la Tournelle, which specialized in wine; and Port Saint-Bernard, which handled wood—and others downstream—such as Port Saint-Nicolas, which specialized in grains and transatlantic shipments.

8. The boats from upstream were broken up and sold as wood for heating, an operation that was originally performed at the bottom of Pont de la Tournelle but was later moved to Île des Cygnes.

9. The three Parisian canals—Ourcq, Saint-Denis, and Saint-Martin—would be constructed by a contractor who then managed them until the city assumed control of them during the Second Empire. Paris's canals then became part of the network of canals that engineers of the École Nationale des Ponts et Chaussées were at that time building in the north of France—an integration that confirmed the Seine's new national role.

10. See "Mémoire des teinturiers de la rue de la Pelleterie à Monsieur de Calonne," dated Sept. 1786, ANP, H 2167; and "Délibération du bureau de la ville," dated Nov. 15, 1786, ANP, H 2168.

11. See Jean-Claude Perrot, *Genèse d'une ville moderne: Caen au XVIIIe siècle* (Paris: Mouton, EHESS, 1975).

12. Robert Coustet and Marc Saboya, *Bordeaux, le temps de l'histoire: Architecture et urbanisme au XIXe siècle (1800–1914)* (Bordeaux: Mollat, 1999).

13. *Toulouse: la ville au XVIIIe siècle*, exhibition catalogue (Toulouse: Conseil général de la Haute-Garonne, 1985).

14. Pierre Lelièvre, *Nantes au XVIIIe siècle: Urbanisme et architecture* (Paris: Picard, 1988).

15. "Mémoire sur le projet de rendre Paris port de mer, et d'y faire arriver des vaisseaux de Rouen dès la première année de l'entreprise," by Passemant, ingénieur du Roy [royal engineer], and Bellart, avocat au Conseil [attorney to the Council], ANP, O¹1693.

16. "Lettres patentes du roi, en forme de déclaration, portant établissement, dans la ville de Paris, d'une nouvelle Halle aux blés, et d'une gare pour les bateaux," dated Nov. 25, 1762, ANP, ADXVI 10.

17. "Affaires de Paris, construction d'une gare sur la Seine, 1762–1768: Requête du procureur général du roi," dated Apr. 23, 1765, Bibliothèque Nationale de France (hereafter BNF), Joly de Fleury, 1425, fol. 31.

18. Bibliothèque de l'École Nationale des Ponts et Chaussées, Ms 2083, dated Dec. 1790.

19. "Mémoire de Mr Bourbon de la Cour, avocat au Parlement, résidant à Saint Dizier," dated May 16, 1765, and "Plan figuré d'un nouveau port qu'il conviendroit placer

aux prés aux Clercs à Paris, Claude Henry, géomètre," n.d., BNF, Joly de Fleury, 1425, fols. 44–45, 46.

20. "Observations sur l'etablissement d'une gare artificielle," n.d., BNF, Joly de Fleury, 1425, fols. 47–48.

21. "Arrest du parlement du 4 septembre 1767," BNF, Joly de Fleury, 1425, fol. 62.

22. "Lettre du roi à Maupéou" dated May 23, 1768, BNF, Joly de Fleury, 1425, fol. 172.

23. See, for example, "Plan de la ville de Paris et de ses faubourgs," B. Jaillot, dated 1770, ANP, NII Seine 75; "Plan de Paris par Lattré," dated 1783, BNF, GeC 3363; and "Plan de Paris dans l'enceinte fiscale des fermiers généraux," dated 1788, ANP, NIII Seine 874.

24. Charles De Wailly, "Projet d'utilité et d'embellissement pour la ville de Paris qui s'accorde avec les projets déjà arrêtés par le gouvernement dans lequel on a rassemblé de nouveaux monuments, des places publiques, des percées nécessaires pour donner des communications, augmenter les courants d'air, et où l'on propose la réunion des trois îles en une seule, les moyens de diriger le courant du bras septentrional de la rivière de manière à le rendre plus navigable, et de l'autre à en former un port ou garre, au centre de la capitale" (BNF, Cartes et plans, GeC 4384). The plan itself is not dated, but a date of 1789 is given in M. Mosser and D. Rabreau, *Charles De Wailly: Peintre architecte dans l'Europe des Lumières* (Paris: Caisse nationale des monuments historiques et des sites, 1979), 71.

25. Bernard Poyet, *Mémoire sur la nécessité de transférer et reconstruire l'Hôtel-Dieu de Paris suivi d'un projet de translation de cet hôpital . . .* (Paris, 1785). Poyet's collaborator on this proposal, which was ultimately rejected, was his fellow architect Claude-Philibert Coqueau.

26. Charles De Wailly, "La halle au blé étant reconnue insuffisante pour les approvisionnements de Paris, on a un devoir d'y supléer en établissant au bord de la rivière après l'isle des Cygnes, un cirque qui simétriseroit avec l'Hôtel-Dieu projeté, et où les bateaux pouroient entrer pour y rester chargés pendant l'hyver, si les magasins étoient assés remplis. On se serviroit aussi l'été de ce bassin pour y faire des joutes et apprendre à nager. Le soubassement du cirque au niveau de la rivière serviroit de bains publics" (1789).

27. Pierre Giraud, "La réunion des trois isles, ou projet d'un canal de navigation sur la branche de rivière du pont Rouge, avec promenade publique; d'un pareil canal à la pointe de l'isle Saint-Louis, servant d'École de natation; d'une gare au centre de la capitale; d'un grand pont, à colonnes, sans avant-bec, et à trois arches, en face du boulevard de l'hôpital général; d'un Château-d'eau, et d'une machine hydraulique, propre à remplacer les pompes Notre-Dame et de la Samaritaine, de quatre ponts de pierre et de brique, à l'alignement des quais; de cinq ponts de fer à une seule arche, dont deux en chaines, pour l'usage des gens de pied seulement; d'un séminaire et d'une communauté de prêtres attachés à l'église paroissiale-métropolitaine, d'un hospice pour les sections Henri IV, de Notre-Dame, et de l'Isle Saint-Louis; d'une Halle, avec logemens au-dessus; de plusieurs magasins, ports, quais, trottoirs et autres constructions de la même importance" (dated 1791, Bibliothèque Historique de la Ville de Paris).

28. To their plans for the Seine can be added an anonymous six-page design dated

1791 and titled "Projet de réunion de l'île Louviers et de l'île Saint-Louis avec l'île de la Cité: Etablissement d'une gare et construction de moulins au centre de la ville, par M.P.S.O.D.L.R." (Bibliothèque de Zurich, 5058) and ideas advanced by "Citizen Belleville in Year 8" (1800) in a document titled "Projet de garre à établir à Paris pour la sureté du commerce pendant toute l'année et particulièrement pendant l'hiver" (ANP, F[14]608[1]).

29. Charles-Louis Corbet and Charles Mangin, "Projet d'une place publique à la gloire de Louis XVI sur l'emplacement de la Bastille, ses fossés et dépendances avec la continuation du rempart jusqu'à la rivière, sur partie des fossés de l'enceinte de l'Arsenal" (dated 1789, ANP, N IV Seine 87, fol. 5).

30. "Projet d'une place sur l'emplacement de la Bastille avec une colonne au centre semblable à celle de Trajan à Rome avec celui d'une gare" (ANP, N III Seine 762/1–3). See also Étienne-Louis-Denis Cathala, "Projet de gare, de pont, de greniers à bleds, et d'une place sur les terrains de la Bastille" (Paris: L'Imprimerie de la veuve Hérissant, 1790).

31. J. C. Périer, *Mémoire sur l'établissement d'une gare à Paris* (Paris: L'Imprimerie Royale, 1790), 6.

32. "Projet de gare du citoyen Périer," n.d., ANP, F[14]608[1].

33. Pierre-François Palloy, "Projet général d'un monument à élever à la gloire de la Liberté, sur les terrains de la Bastille, Isle Louviers et dépendances, dédié à la Nation, et présenté à l'Assemblée nationale le 11 mars 1792"; and "Place circulaire avec un large bassin de commerce à mi-chemin du fleuve, Halles aux boissons, pont vers le jardin des Plantes," ANP, F[14]10078[1]/18. The name "jardin des plantes" dates the project to a period before the June 1793 decree that transformed the garden into the Muséum national d'Histoire naturelle.

34. Le Clerc, "Plan d'un bassin de commerce et de garre et d'une halle aux boissons sur l'emplacement des jardins de l'arsenal, jardins des ci-devant Célestins et isle Louviers d'une grande place publique avec douze rues convergentes à son centre, sur le terrain de la Bastille et de ses fossés; d'un monument à la gloire des armées françaises et occupant le centre de la place, enfin d'un pont établissant communication entre les faubourgs Antoine et Marceau," dated Vendémiaire, an VII (Sept.–Oct. 1798), ANP, F[14]6729.

35. "Mémoire sur l'établissement d'un bassin de commerce et de gare à Paris sur les terrains des fossés, jardins, cours et bâtiments des ci-devants Célestins, portion de l'Isle Louviers et propriétés particulières au delà de la Contrescarpe," dated 15 Brumaire, an VII (Nov. 5, 1798), ANP, F[14]6729.

36. Jean-Pierre Brullée, "Mémoire sur les moyens de former des établissements d'utilité publique, et d'en assurer les propriétés au profit de la Nation, sans être à charge au trésor national," dated 20 Brumaire, an VIII (Nov. 11, 1799), ANP, AD XIII 13.

37. "Arrêté qui détermine les fonctions du préfet de police," dated 12 Messidor, an VIII (July 1, 1800), ANP, AD XVI 71.

38. The order of 1 Messidor, an XI (June 21, 1803), divided the Basin of the Seine into nine boroughs. The fifth, from Choisy to Pecq, included Paris. This division shows the preoccupation with coherence between boroughs; the capital was only one link among others. This longitudinal view diminishes the city's importance by making it merely one segment of the Seine.

39. See Bill Luckin, *Pollution and Control: A Social History of the Thames in the Nineteenth Century* (Bristol: Adam Hilger, 1986); and Philippe Cebron de Lisle, *L'eau à Paris au XIXe siècle* (Paris: Association générale des hygiénistes et techniciens municipaux, 1991).

40. "Loi relative à la navigation intérieure," dated May 31, 1846, *Bulletin des lois du royaume de France*, ser. 9, 32 (July 1846): 389.

41. I return to this question in "Paris sous les eaux: La grande crue de 1910," *L'Histoire* 257 (Sept. 2001): 46–49.

42. For full details of recently proposed and/or implemented flood prevention measures, see the Web site for France's Ministry of Ecology and Sustainable Development, http://www.environnement.gouv.fr/ile-de-france.

43. In 1992, UNESCO added the banks of the Seine to its list of "World Heritage" sites; see http://whc.unesco.org/en/list/600. Founded in 1867 on the occasion of the Exposition Universelle, the Compagnie des Bateaux Omnibus still transports travelers between the Pont Napoléon (Pont National) and the Auteuil Viaduct. See M. Merger, "Les bateaux mouches (1867–1914)," in *Métropolitain: L'autre dimension de la ville* (Paris: Catalogue de l'exposition de la Bibliothèque Historique de la Ville de Paris, 1988), 20–21.

44. Jean Pelletier, "De l'économie au ludique et au paysagement urbain," in *La ville et le fleuve*, ed. Comité des travaux historiques et scientifiques (Paris: CTHS, 1989).

45. Jacques Béthemont, *Les grands fleuves, entre nature et société*, 2nd ed. (Paris: Armand Colin, 2000).

Chapter 4: Pittsburgh's Three Rivers

1. Edward K. Muller, "The Point," in *Geographical Snapshots of North America*, ed. Donald G. Janelle (New York: Guilford Press, 1992), 231–34.

2. David J. Cuff et al., *The Atlas of Pennsylvania* (Philadelphia: Temple University Press, 1989), 52–67. Some have questioned whether there are in fact three rivers, arguing that the Allegheny and the Ohio are one river, which the Monongahela joins. However, ever since Europeans first explored this region, most people have viewed these as three separate rivers.

3. Paul A. W. Wallace, *Indians in Pennsylvania* (Harrisburg: Pennsylvania Historical and Museum Commission, 1981).

4. Leland D. Baldwin, *Pittsburgh: The Story of a City, 1750–1865* (Pittsburgh, PA: University of Pittsburgh Press, 1937).

5. Shera A. Moxley, *From Rivers to Lakes: Engineering Pittsburgh's Three Rivers*, 3 Rivers 2nd Nature History Report (Pittsburgh, PA: STUDIO for Creative Inquiry, Carnegie Mellon University, 2001), 18, available at http://3r2n.cfa.cmu.edu/history/engineer/index.htm; Leland R. Johnson, *The Headwaters District: A History of the Pittsburgh District, U.S. Army Corps of Engineers* (Pittsburgh, PA: Pittsburgh District, U.S. Army Corps of Engineers, 1978), 1–101.

6. Catherine Elizabeth Reiser, *Pittsburgh's Commercial Development, 1800–1850* (Harrisburg: Pennsylvania Historical and Museum Commission, 1951); Richard C. Wade, *The Urban Frontier: Pioneer Life in Early Pittsburgh, Cincinnati, Lexington, Louisville, and St.

Louis (Chicago: University of Chicago Press, 1964), 39–46. Wade's study was originally issued in 1959 by Harvard University Press under the title *The Urban Frontier: The Rise of Western Cities, 1790–1830.*

7. Edward K. Muller, "River City," in *Devastation and Renewal: An Environmental History of Pittsburgh and Its Region,* ed. Joel A. Tarr (Pittsburgh, PA: University of Pittsburgh Press, 2003), 41–63, esp. 45–51.

8. John. N. Ingham, *Making Iron and Steel: Independent Mills in Pittsburgh, 1820–1920* (Columbus: Ohio State University Press, 1991).

9. Moxley, *From Rivers to Lakes,* 10–25; Leland R. Johnson, *The Davis Island Lock and Dam, 1870–1922* (Pittsburgh, PA: U.S. Army Corps of Engineers, 1985).

10. Muller, "River City," 51–56; Edward K. Muller and Joel A. Tarr, "The Interaction of Natural and Built Environments in the Pittsburgh Landscape," in *Devastation and Renewal: An Environmental History of Pittsburgh and Its Region,* ed. Joel A. Tarr (Pittsburgh, PA: University of Pittsburgh Press, 2003), 11–40.

11. Quoted in Joel A. Tarr and Terry F. Yosie, "Critical Decisions in Pittsburgh Water and Wastewater Treatment," in *Devastation and Renewal: An Environmental History of Pittsburgh and Its Region,* ed. Joel A. Tarr (Pittsburgh, PA: University of Pittsburgh Press, 2003), 64–88, esp. 76.

12. Ibid.

13. Erwin E. Lanpher and C. F. Drake, *City of Pittsburgh, Pennsylvania: Its Water Works and Typhoid Fever Statistics* ([n.p.], 1930), 23–25. An online edition of this work, published in 1999 by the University of Pittsburgh Digital Research Library, is available at http://digital.library.pitt.edu/cgi-bin/t/text/text-idx?idno=00abv4381m;view=toc;c=pitttext.

14. Tarr and Yosie, "Critical Decisions," 69.

15. Ibid., 70–72.

16. Ibid., 70–74. Allegheny City was a separate city directly across the Allegheny River from Pittsburgh until annexed by its larger neighbor in 1907.

17. Ibid., 74–77. Hazen had been the key engineer on the Pittsburgh Filtration Commission.

18. Ibid., 76–77.

19. Nicholas Casner, "Acid Mine Drainage and Pittsburgh's Water Quality," in *Devastation and Renewal: An Environmental History of Pittsburgh and Its Region,* ed. Joel A. Tarr (Pittsburgh, PA: University of Pittsburgh Press, 2003), 89–109.

20. Ibid.

21. Pittsburgh Flood Commission, *Report of the Flood Commission of Pittsburgh, Penna. Containing the results of the surveys, investigations and studies made by the Commission for the purpose of determining the causes of, damage by and methods of relief from floods in the Allegheny, Monongahela, and Ohio rivers at Pittsburgh, Penna., together with the benefits to navigation, sanitation, water supply and water power to be obtained by river regulation* (Pittsburgh, PA: Murdoch, Kerr & Co., 1912); Roland M. Smith, "The Politics of Pittsburgh Flood Control, 1908–1936," *Pennsylvania History* 42 (1975): 5–24.

22. Frederick Law Olmsted Jr., *Main Thoroughfares and the Down Town District*

(Pittsburgh, PA: Pittsburgh Civic Commission, 1911), 19–30, also available at http://pghbridges.com/articles/olmsted/00intro/olm000a.htm; John F. Bauman and Edward K. Muller, *Before Renaissance: Planning in Pittsburgh, 1889–1943* (Pittsburgh, PA: University of Pittsburgh Press, 2006), 72–86.

23. Citizens Committee on the City Plan of Pittsburgh, "Waterways: A Part of the Pittsburgh Plan," Report No. 6 (Pittsburgh, PA: Citizens Committee on the City Plan of Pittsburgh, 1923), 13–15, esp. 13; also available at http://digital.library.pitt.edu/cgi-bin/t/text/text-idx?idno=00hc02396m;view=toc;c=pitttext.

24. Tarr and Yosie, "Critical Decisions," 78–79.

25. Ibid., 77–86.

26. Roy Lubove, *Twentieth-Century Pittsburgh*, 2nd ed., 2 vols. (Pittsburgh, PA: University of Pittsburgh Press, 1996), 1:106–41; Shelby Stewman and Joel A. Tarr, "Four Decades of Public-Private Partnerships in Pittsburgh," in *Public-Private Partnerships in American Cities*, ed. R. Scott Fosler and Renee A. Berger (Lexington, MA: Lexington Books, 1982), 59–127; Sherie R. Mershon, "Corporate Social Responsibility and Urban Revitalization: The Allegheny Conference on Community Development, 1943–1968" (Ph.D. diss., Carnegie Mellon University, 2000); Roland M. Smith, "The Politics of Pittsburgh Flood Control, 1936–1960," *Pennsylvania History* 44 (1977): 3–24; Robert C. Alberts, *The Shaping of the Point: Pittsburgh's Renaissance Park* (Pittsburgh, PA: University of Pittsburgh Press, 1980).

27. Muller, "River City," 57–63.

28. Lubove, *Twentieth-Century Pittsburgh*, 2:201–2.

29. M. Graham Netting, *50 Years of the Western Pennsylvania Conservancy* (Pittsburgh, PA: Western Pennsylvania Conservancy, 1982), 19.

30. Griswold, Winters, and Swain, *A Master Plan for the Development of Riverfronts and Hillsides in the City of Pittsburgh: An Analysis of Their Best Possible Uses, for the Enhancement of the City, and the Enjoyment of Its Citizens* (Pittsburgh, PA: Department of Parks and Recreation, 1959), 3–5, esp. the dedication.

31. Edgar M. Hoover, *Economic Study of the Pittsburgh Region*, 3 vols. (Pittsburgh, PA: University of Pittsburgh Press, 1963); John P. Hoerr, *And the Wolf Finally Came: The Decline of the American Steel Industry* (Pittsburgh, PA: University of Pittsburgh Press, 1988).

32. Lubove, *Twentieth-Century Pittsburgh*, 2:5–23.

33. For more on this project, see Chan Krieger & Associates et al., *A Vision Plan for Pittsburgh's Riverfronts* ([Pittsburgh, PA]: Riverlife Task Force, [2001]), also available at http://www.riverlifetaskforce.org/about/news/publications.

34. Muller, "River City," 62–63.

35. Robert Hoskin, Michael Koryak, and Linda Stafford, *Fishes of Small Tributaries to the Allegheny and Monongahela Rivers in Allegheny County, Pennsylvania*, 3 Rivers 2nd Nature Fish Report: Monongahela and Allegheny (Pittsburgh, PA: STUDIO for Creative Inquiry, Carnegie Mellon University, 2002), pt. III, sec. A, also available at http://3r2n.cfa.cmu.edu/water/fish/monAlleg/index.htm.

36. Robert Hoskin, Michael Koryak, et al., "The Impact of Above Grade Sewerline

Crossing on the Distribution and Abundance of Fishes in Recovering Small Urban Streams of the Upper Ohio River Valley," *Journal of Freshwater Ecology* 16.4 (2001): 591–98, esp. 594.

37. Susan Kalisz and Jessica Dunn, *Riverbank Vegetation*, 3 Rivers 2nd Nature Botany Report: Allegheny River (Pittsburgh, PA: STUDIO for Creative Inquiry, Carnegie Mellon University 2002), pt. IV, also available at http://3r2n.cfa.cmu.edu/land/bot/alleg/index.htm. See also Pennsylvania Economy League, *Investing in Clean Water: A Report from the Southwestern Pennsylvania Water and Sewer Infrastructure Project Steering Committee* (Pittsburgh, PA: Southwestern Pennsylvania Water and Sewer Infrastructure Project, 2002). Attempts to resolve the mine acid problem included state and federal mine-sealing programs, but these achieved only limited success. Not until the passage of state and federal legislation requiring active mining operations to treat polluted water prior to discharge did conditions finally improve; see Casner, "Acid Mine Drainage." In addition, see Jan Ackerman, "Mount Washington Residents Question Development Plans," *Pittsburgh Post Gazette*, May 29, 2003, http://www.post-gazett.com/neigh_city/20030529hillsidec5.asp; and Tom Barnes, "Council OKs Mining Project as Precursor to Hays Development," *Pittsburgh Post Gazette*, July 31, 2003, http://www.post-gazette.com/neigh_city20030731miningc2.asp.

38. United States, *Nine Mile Run Ecosystem Restoration Project* ([Pittsburgh, Pa.]: U.S. Army Corps of Engineers, Pittsburgh District, 2000); see also http://www.lrp.us ace.army.mil/pm/9mile.htm.

39. For more information on these projects, see the U.S. Army Corps of Engineers' Web site for information on the *Ohio River Mainstem Systems Study (ORMSS) Interim Feasibility Report: Ohio River Ecosystem Restoration Program*, 7 vols. (2000), at http://www.lrl.usace.army.mil/ORMSS/; the U.S. Fish and Wildlife Service's Web site on its Ohio River Islands National Wildlife Refuge, at http://northeast.fws.gov/planning/ORI_WEB/chap1.htm; and the Allegheny Land Trust's Web site, especially the pages related to the Whetzel Preserve, at http://www.alleghenylandtrust.org/properties/whetzel/overview/index.html.

40. Dieter Schott, "Rivers as Urban Environments: Cycles and Constellations in the Management and Perception of British Cities," lecture delivered at "The Making of European Contemporary Cities: An Environmental History," Third International Round Table on Urban Environmental History of the 19th and 20th Century, University of Siena, June 24–27, 2004, 89–96; Isabelle Backouche, "From Parisian River to National Waterway: The Social Functions of the Seine, 1750–1850" (chapter 3 in this volume).

41. For a comparison of rural and urban river development, see Susan Q. Stranahan, *Susquehanna: River of Dreams* (Baltimore: Johns Hopkins University Press, 1993); and Ari Kelman, *A River and Its City: The Nature of Landscape in New Orleans* (Berkeley: University of California Press, 2003). For an extreme example of an urban river, see Blake Gumprecht, *The Los Angeles River: Its Life, Death, and Possible Rebirth* (Baltimore: Johns Hopkins University Press, 1999).

42. Eva Jakobsson, "How Do Historians of Technology and Environmental Historians Conceive the Harnessed River?" lecture delivered at "Rivers in History: Designing

and Conceiving Waterways in Europe and North America" conference, Dec. 4–7, 2003, German Historical Institute, Washington, DC.

43. Edward K. Muller, "The Legacy of Industrial Rivers," *Pittsburgh History* 72 (1989): 46–75.

44. William Cronon, "Introduction: In Search of Nature," in *Uncommon Ground: Toward Reinventing Nature,* ed. William Cronon (New York: Norton, 1995), 23–66, esp. 25.

Chapter 5: The Cultural and Hydrological Development of the Mississippi and Volga Rivers

1. In *Seeing Like a State: How Certain Schemes to Improve the Human Condition Have Failed* (New Haven: Yale University Press, 1998), James C. Scott uses the term "high modernism" to describe the ideology that promoted and informed large-scale, state-supported engineering projects. Scott contends that this ideology "could be found across the political spectrum from left to right [especially] among those who wanted to use state power to bring about huge, utopian changes in people's work habits, living patterns, moral conduct, and worldview" (5).

2. William Patrick O'Brien et al., *Gateways to Commerce: The U.S. Army Corps of Engineers' 9-Foot Channel Project on the Upper Mississippi River* (Denver: National Park Service, 1992), 14.

3. David E. Lilienthal, *TVA: Democracy on the March* (New York: Harper and Row, 1944); Sheila Fitzpatrick, *The Russian Revolution* (Oxford: Oxford University Press, 1994), 149.

4. Scott, *Seeing Like a State,* 4. Complementing Scott's work is Paul R. Josephson's *Industrialized Nature: Brute Force Technology and the Transformation of the Natural World* (Washington, DC: Island Press, 2002). This excellent book and its theme of "brute force technologies" is an invaluable resource for an understanding of Soviet attitudes toward nature. See also Deborah Fitzgerald, *Every Farm a Factory: The Industrial Ideal in American Agriculture* (New Haven: Yale University Press, 2003).

5. Mark Twain, *Life on the Mississippi* (1883; rpt., New York: Signet Classic, 1961), 1.

6. Walter Havighurst, *Voices on the River: The Story of the Mississippi Waterways* (Minneapolis: University of Minnesota Press, 1964), 25–28.

7. "The Great Railroad Excursion," *New York Daily Times (1851–1857),* June 14, 1854, 2. This account, one of several from a correspondent identified only by the single initial "W" is reprinted in Nancy Goodman and Robert Goodman, *Paddlewheels on the Upper Mississippi, 1823–1854: How Steamboats Promoted Commerce and Settlement in the West* (Stillwater, MN: University Washington County Historical Society, 2003), 2–3, 6–9.

8. Goodman and Goodman, *Paddlewheels,* 10.

9. See, e.g., Twain, *Life on the Mississippi,* 59. In addition to Twain's, there are numerous early accounts of the Mississippi River from such explorers as Zebulon Pike, James Duane Doty, and Stephen Long.

10. Twain, *Life on the Mississippi,* 2.

11. Sam J. Graber, "The Upper Mississippi River Improvement Association" (M.A.

thesis, University of Wisconsin–La Crosse, 1968), 17. Additional information regarding engineering developments and barge traffic on the Upper Mississippi River comes from an interview with Army Corps of Engineers staff at Lock No. 7, Apr. 28, 2003; James G. Wiener et al., "Mississippi River," in *Status and Trends of the Nation's Biological Resources*, ed. Michael J. Mac et al., 2 vols. (Fort Collins, CO: U.S. Dept. of the Interior, U.S. Geological Survey, 1998), http://biology.usgs.gov/s+t/SNT/; Richard Hoops, *A River of Grain—The Evolution of Commercial Navigation on the Upper Mississippi River* (Madison: University of Wisconsin–Madison College of Agriculture and Life Sciences Research Report, 1993); and John O. Anfinson, *The River We Have Wrought: A History of the Upper Mississippi* (Minneapolis: University of Minnesota Press, 2003).

12. Christopher Ely, *This Meager Nature: Landscape and National Identity in Imperial Russia* (DeKalb: Northern Illinois University Press, 2002), 76 (first quote), 35 (second quote).

13. See N. A. Nekrasov, "On the Volga River," *Lirika [Lyrics]* (Moscow: Detskaja Literature, 1976), 81–89.

14. George Heard Hamilton, *The Art and Architecture of Russia*, 3rd ed. (New Haven: Yale University Press, 1983); *Isaak Levitan* (Moscow: White City, 2000); and Orlando Fieges, *Natasha's Dance: A Cultural History of Russia* (New York: Henry Holt, 2002).

15. Nekrasov, "On the Volga River," 89.

16. This information about the history of the Moscow-Volga Canal is drawn from interviews I conducted with Professor E. Y. Shimon on Sept. 6, 2003, and with N. Fedorov on Sept. 10, 2003; from N. Fedorov, *Has the Minister a Car? Stories about the Moscow-Volga Canal Construction* (Dimitrov: SPAS, 1997); and from Romuald Khokhlov, "The Inauguration of the Moscow-Volga Canal," *Ploshchad Mira*, June 10, 1997. For general works on the Volga and the Moscow-Volga Canal, I have consulted Josephson, *Industrialized Nature;* Douglas Wiener, *A Little Corner of Freedom: Russian Nature Protection from Stalin to Gorbachev* (Berkeley: University of California Press, 1999); David R. Shearer, *Industry, State, and Society in Stalin's Russia: 1926–1934* (Ithaca, NY: Cornell University Press, 1996); and Marq de Villiers, *Down the Volga: A Journey through Mother Russia in a Time of Troubles* (New York: Viking, 1992).

17. In two excellent works on the Upper Mississippi River valley, the authors both stress how the river has become a human design or, in Phil Scarpino's words, a "human artifact." See Phillip V. Scarpino, *Great River: An Environmental History of the Upper Mississippi, 1850–1950* (Columbia: University of Missouri Press, 1985). John Anfinson treats the same theme, as evidenced in his title *The River We Have Wrought*.

18. Graber, "The Upper Mississippi River Improvement Association," 43. In 1940, 2.4 million tons of goods were shipped along the Mississippi; by 2000 that figure had risen to 83 million tons.

19. Ibid., 62.

20. Several excellent accounts of the history of navigation on the Upper Mississippi River are in print, including Anfinson, *The River We Have Wrought;* C. R. Fremling and T. O. Claflin, "Ecological History of the Upper Mississippi River," in *Contaminants in the Upper Mississippi River: Proceedings of the 15th Annual Meeting of the Mississippi River*

Research Consortium, ed. James G. Wiener et al. (Boston: Butterworth, 1984), 5–25; and Hoops, *A River of Grain*. The Oral History Collection at the University of Wisconsin–La Crosse includes many firsthand accounts by residents who have lived along the river and witnessed its development.

21. O'Brien et al., *Gateways to Commerce*, 15. Water projects in the United States have a long history of visual and print culture, particularly in the arid American West. Documentation of water projects became standard practice with the passage of the Reclamation Act of 1902 and subsequent creation of the U.S. Reclamation Service.

22. According to one U.S. government publication, "Hoover Dam was perhaps the most significant American public-works project of the twentieth century. It was the first large Federal conservation undertaking based on multiple-purpose objectives" (quoted in Michael C. Robinson, *Water for the West: The Bureau of Reclamation, 1902–1977* [Chicago: Public Works Historical Society, 1979], 51). An interesting discussion on the Tennessee Valley Authority and FDR's commitment to modernize and electrify American homes can be found in Ronald C. Tobey, *Technology as Freedom: The New Deal and the Electrical Modernization of the American Home* (Berkeley: University of California Press, 1996). For additional insights into Stalin's treatment of Soviet engineers, see Loren R. Graham, *The Ghost of the Executed Engineer: Technology and the Fall of the Soviet Union* (Cambridge, MA: Harvard University Press, 1993).

23. For a discussion of the growing dominance of the expert, see Samuel P. Hays, *Conservation and the Gospel of Efficiency: The Progressive Conservation Movement, 1890–1920* (1959; rpt., Cambridge, MA: Harvard University Press, 1975). On Elwood Mead, who exemplified the expert and shaped water policy in the United States and Australia, see James R. Kluger, *Turning on Water with a Shovel: The Career of Elwood Mead* (Albuquerque: University of New Mexico Press, 1992). The Leon Trotsky quote is from C. Wright Mills, *The Marxists* (Hammondsworth: Penguin Books, 1963), 278–79.

24. Franklin Delano Roosevelt, quoted in Tamberlain Jacobs, "Lake City Marina: From Pond to Prosperity," in *Big River Reader: An Anthology of Stories about the Upper Mississippi, from the First Four Years of Big River*, ed. Pamela Eyden et al. (Winona, MN: Big River, 1996), 136.

25. David E. Lilienthal, *TVA: Democracy on the March* (New York: Harper and Row, 1944), 212.

26. See Tobey, *Technology as Freedom*.

27. Joseph Stalin, quoted in Fitzpatrick, *The Russian Revolution*, 130.

28. Douglas Wiener, *A Little Corner of Freedom*, 355; Fedorov interview; Fedorov, *Has the Minister a Car?*; Khokhlov, "The Inauguration of the 'Moscow-Volga' Canal."

29. Fedorov interview; Fedorov, *Has the Minister a Car?*; Khokhlov, "The Inauguration of the 'Moscow-Volga' Canal." For an excellent discussion of the high-level Soviet officials surrounding Stalin, such as Yagoda, see Simon Sebag Montefiore, *Stalin: The Court of the Red Tsar* (New York: Knopf, 2004).

30. Anne Applebaum, *Gulag: A History* (New York: Anchor Books, 2003), 55–56. Applebaum's exhaustively researched study provides ample evidence of the central role that forced labor played in Russia's modernization. Khokhlov, "The Inauguration of the

'Moscow-Volga' Canal"; Fedorov, *Has the Minister a Car?*; Two small museums in Dubna and Dimitrov preserve news clippings and photographs of the canal's construction. The Dimitrov museum also has a few copies of Kun's work.

31. Khokhlov, "The Inauguration of the 'Moscow-Volga' Canal"; interview with Nikolai Preclonov, Sept. 24, 2003. In this interview, conducted in Dimitrov, Russia, Preclonov asserted that the Moscow-Volga Canal "civilized this part of the country."

32. For a concise discussion of contemporary environmental problems in the Upper Mississippi River basin, see James G. Wiener et al., "Mississippi River." For a more detailed discussion, see James G. Wiener et al., eds., *Contaminants in the Upper Mississippi River: Proceedings of the 15th Annual Meeting of the Mississippi River Research Consortium* (Boston: Butterworth, 1984). For annual lists of America's "Most Endangered Rivers" from 1986 to the present, see the American Rivers Web site, http:// www.americanrivers.org.

33. Wiener, et al., *Contaminants in the Upper Mississippi River.*

34. Information given here on the environmental concerns of the Volga River comes from several papers presented at the Second International Conference on Great Rivers as Attractors for Local Civilizations, held at Assiut University, Egypt, Oct. 12–14, 2003. A volume of the proceedings of the conference is forthcoming.

35. Patrick McCully, *Silenced Rivers: The Ecology and Politics of Large Dams* (London: Zed Books, 2001), 152; Scott, *Seeing Like a State.*

36. Anfinson, *The River We Have Wrought,* 279. In a pioneering report produced by Hiram Chittenden in 1899 for the U.S. Department of Agriculture, when considering the construction of reservoirs in the American West Chittenden recommended that the federal government assume responsibility for the building of reservoir sites for a host of reasons. He discussed the potential problems arising from the interstate nature of many of the rivers and stated that since the government owned much of the land to be reclaimed, federal sponsorship of irrigation systems represented sound business practices. See U.S. Department of Agriculture, Office of Experimental Stations, *The Use of Water in Irrigation: Report of Investigations Made in 1899,* Bulletin No. 8 (Washington, DC, 1900). For a thorough examination of the federal government's role in national water policy, see Donald J. Pisani, *Water and American Government: The Reclamation Bureau, National Water Policy, and the West, 1902–1935* (Berkeley: University of California Press, 2002).

37. The remark about "limitless development" is attributed to an unnamed Soviet scientist in the 1930s and is quoted in Ramachandra Guha, *Environmentalism: A Global History* (New York: Longman, 2000), 126; Tobey, *Technology as Freedom.*

38. Josephson, *Industrialized Nature,* 3.

39. Jawaharlal Nehru, quoted in McCully, *Silenced Rivers,* 20. Literature on water development and modernization is rich with essays by eminent authors such as Arundhati Roy, Marq de Villiers, Vandana Shiva, Maude Barlow, and Tony Clarke, to name just a few. One of the best-known recent articles on dam building and modernization is Nick Cullather, "Damming Afghanistan: Modernization in a Buffer State," *Journal of American History* 89.2 (2002): 512–37.

Chapter 6: River Diking and Reclamation in the Alpine Piedmont

1. Jean-Paul Bravard, "La métamorphose des rivières des Alpes françaises à la fin du Moyen-âge et à l'époque moderne," *Bulletin de la Société Géographique de Liège* 25 (1989): 145–57; Pierre-Gil Salvador, "La métamorphose des cours du Drac et de l'Isère à l'époque moderne dans la région grenobloise (Isère, France)," *Physio-Géo* 22/23 (1991): 173–78; Denis Coeur, "Genesis of a Public Policy for Flood Management in France: The Case of the Grenoble Valley (XVIIth–XIXth Centuries)," in *Palaeofloods, Historical Data & Climatic Variability: Applications in Flood Risk Assessment*, ed. Varyl R. Thorndycraft et al., Proceedings of the PHEFRA International Workshop, Barcelona, Oct. 16–19, 2002, 373–78.

2. This map, which is known as the "Carte du Duc de Berwick," was designed by Roussel in 1711 and is now in the collection of the Bibliothèque Nationale de France, Paris (shelfmark Ge.A. 1073).

3. *Mappe sarde*, Archives Départementales de la Savoie, Chambéry (série C); *Cadastre napoléonien*, Archives Départementales de la Savoie, Chambéry (série L, nos. 1007, 1008, 1009). See also Jacky Girel et al., "Landscape Structure and Historical Processes along Diked European Valleys: A Case Study of the Arc/Isère Confluence (Savoie, France)," *Environmental Management* 21 (1997): 891–907; and Jacky Girel et al., "Biodiversity and Land-Use History of the Alpine Riparian Landscapes (the Example of the Isère River, France)," in *Multifunctional Landscapes*, vol. 3, *Continuity and Change*, ed. Ülo Mander and Marc Antrop (Southampton, England: WIT, 2004), 167–200.

4. Jean-Baptiste Duvergier, "Loi relative au dessèchement des marais (16 septembre 1807)," in *Collection complète des lois, décrets, ordonnances, réglements et avis du Conseil d'Etat de 1788 à 1824* (Paris: Guyot and Scribe, 1826), 143–45.

5. Bravard, "La métamorphose des rivières des Alpes françaises"; Darren S. Crook et al., "Forestry and Flooding in the Annecy Petit Lac Catchment, Haute-Savoie 1730–2000," *Environment and History* 8 (2002): 403–28; Emanuela Guidoboni, "Human Factors, Extreme Events, and Floods in the Lower Po Plain (Northern Italy) in the 16th Century," *Environment and History* 4 (1998): 279–308.

6. Jules Charpentier de Cossigny, *Hydraulique agricole: Aménagement des eaux, irrigation des terres, création des prairies, dessèchements, limonage et colmatage, curage, drainage* (Paris: Baudry, 1889).

7. For a detailed description of Garella's map, see Louis Léger, *Rapport fait à l'administration du département du Mont-Blanc sur le projet de canal de l'Isère* (Chambéry: Gorrin and Associates, 1804).

8. Jean-Baptiste Rougier de la Bergerie, *Recherches sur les principaux abus qui s'opposent aux progrès de l'agriculture* (Paris: Buisson, 1788); Jean-Christian-Marc Boudin, *Traité de géographie et de statistiques médicales et des maladies endémiques* (Paris: J.-B. Baillère et fils, 1857).

9. Marc-Antoine Puvis, *Des étangs, de leur construction, de leur production et de leur dessèchement* (Paris: Huzard, 1844); Edouard de Dienne, *Histoire du dessèchement des lacs et des marais en France avant 1789* (Paris: Champion et Guillaumin, 1891).

10. Duvergier, "Loi relative au dessèchement des marais (16 septembre 1807)." See

also Antoine Noly et al., *Mémoire sur le projet de diguement de l'Isére (d'après Garella et fils) à son Excellence le Ministre de l'Intérieur*, Archives Départementales de la Savoie, série L, additions, travaux publics, no. 7 (Chambéry, 1811).

11. Ingénieur Sollier, *Mémoire sur le diguement des torrents et des rivières de la Savoie*, Archives Départementales de la Savoie, série FS, additions, travaux publics, no. 27 (Chambéry, 1816).

12. Instead of stabilizing the border, however, the sudden change in the course of the river caused frequent disputes over ownership of the islands used for grazing and for supplying firewood.

13. Marie-François-Benjamin Daussé, "Etudes relatives aux inondations et à l'endiguement continu dans l'ancien royaume sarde," *Mémoires présentés par divers savants à l'Académie des Sciences de l'Institut de France (Paris)* 20 (1872): 428–63; Jules Guigues, *Mémoire sur le diguement de l'Isère dans la Combe de Savoie* (Grenoble: Allier, 1891).

14. Joseph Mosca, *Rapport sur l'état de situation des atterrissements des délaissés de l'Isère et des travaux de canalisation de ses affluents*, Archives Départementales de la Savoie, série 1-FS, no. 2419 (Chambéry, 1852); Joseph Mosca, *Mémoire sur l'endiguement de l'Isère et de l'Arc en Savoye*, Archives Départementales de la Savoie, série S, no. 762 (Chambéry, 1860).

15. Antoine Replat et al., "Rapport sur les atterrissements de l'Isère et les effets qu'ils produisent," *Gazette de l'Association Agricole de Turin* 24 (1844): 198–200; Antoine Chiron, *Atterrissements artificiels dans la vallée de l'Isère, cause des fièvres endémiques et périodiques de cette vallée et moyens de les prévenir* (Turin: Chirio and Mina, 1846).

16. Vittorio Fossombroni, *Memorie Idraulico-Storiche sopra la Val-di-Chiana* (Florence: Gaetano Cambiagi, 1789).

17. See Benjamin Nadault de Buffon, *Des submersions fertilisantes comprenant les travaux de colmatage, limonage et irrigations d'hiver* (Paris: Dunod, 1867); Heinrich Speck, "Kolmatierungen im Domleschg und deren agrarpedologische Bedeutung," quoted in Toni Pfiffner, "Reservat Munté: Entstehungsgeschichte, Artenvielfalt und Pflege," *Jahres Bericht der Naturforschende Gesellschaft Graubünden* 109 (2000): 125–218; and Victor J. A. Yvart, *Excursion agronomique en Auvergne principalement aux environs des Monts-Dor et du Puy-de-Dôme, suivie de recherches sur l'état et l'importance des irrigations en France* (Paris: L'Imprimerie Royale, 1819).

18. Jacky Girel, "Old Distribution Procedures of Both Water and Matter Fluxes in Floodplains of Western Europe: Impact on Present Vegetation," *Environmental Management* 18 (1994): 203–21.

19. Chiron, *Atterrissements artificiels dans la vallée de l'Isère;* and Joseph Mosca, "Travaux d'assainissement en Savoie," *Gazette de l'Association Agricole de Turin* 44 (1844): 329–32; Antoine Drizard, "Mémoire sur le colmatage des terrains de la vallée de l'Isère, partie comprise dans la Savoie, entre le pont de Grésy et la limite du département," *Annales des Ponts et Chaussées* 16 (1868): 593–632; and Louis Choron, *Mémoire sur les colmatages de la vallée de l'Isère entre Albertville et la limite du département de la Savoie*, Archives Départementales de la Savoie (Chambéry, 1870), 34 SPc 3, 1–103.

20. During this hundred-year flood the river discharge downstream from the Arc River junction was more than 1,300 m^3 s^{-1} (mean discharge of the river: 460 m^3 s^{-1}) (Archives Départementales de la Savoie [Chambéry], 33 SPc, no. 11).

21. Archives Départementales de la Savoie (Chambéry), 33 SPc, no. 9.

22. Choron, *Mémoire sur les colmatages de la vallée de l'Isère.*

23. Ibid.

24. See Louis-Etienne-François Héricart de Thury, *Du dessèchement des terres cultivables sujettes à être inondées* (Paris: Huzard, 1831); and Hervé Mangon, *Instructions pratiques sur le drainage*, 3rd ed. (Paris: Dunod, 1862).

25. Daniel Zolla, *Code-manuel du propriétaire-agriculteur* (Paris: Giard and Brière, 1894).

26. Jules Guigues, *Des Associations syndicales appliquées à l'Agriculture* (Chambéry: Ménard, 1892).

27. François Gex, "Le diguement de l'Isère dans la Combe de Savoie," *Revue de Géographie Alpine* 28 (1940): 1–71.

28. Most of the diking projects in the Isére valley were completed in the twenty-five-year period from 1829 to 1854, while warping systems were under construction between 1837 and 1890.

29. Although Jean-Paul Haghe focuses on large French coastal marshes in his doctoral dissertation "Les eaux courantes et l'Etat en France 1789–1920: Du contrôle institutionnel à la fétichisation marchande" (EHESS, Paris, 1998), many of his observations and conclusions are equally relevant to the wetlands of the Combe de Savoie.

30. See Daniel Speich, "Draining the Marshlands, Disciplining the Masses: The Linth Valley Hydro Engineering Scheme (1807–1823) and the Genesis of Swiss National Unity," *Environment and History* 8 (2002): 429–47.

31. Jacky Girel and Olivier Manneville, "Present Species Richness of Plant Communities in Alpine Stream Corridors in Relation to Historical River Management," *Biological Conservation* 85.1 (1998): 21–33.

Chapter 7: Holding the Line

1. Mark Cioc, *The Rhine: An Eco-Biography, 1815–2000* (Seattle: University of Washington Press, 2002), 77–99, 109–26; John R. McNeill, *Something New under the Sun: An Environmental History of the Twentieth-Century World* (New York: Norton, 2000), 118–33, 147–48.

2. For the purposes of this essay, I use Sheail's definition of sustainability, which emphasizes managing waterways as a "stock of assets" to be protected for a wide range of users with competing agendas. See John Sheail, "The Sustainable Management of Industrial Watercourses: An English Historical Perspective," *Environmental History* 2 (1997): 197–215, esp. 198. See also John Sheail, "Public Interest and Self-interest: The Disposal of Trade Effluent in Interwar England," *Twentieth Century British History* 4 (1993): 149–70; John Sheail, "'Never Again': Pollution and the Management of Watercourses in Postwar Britain," *Journal of Contemporary History* 33 (1998): 117–33. For Brüggemeier's own argument, see Franz-Josef Brüggemeier, "A Nature Fit for Industry: The Environmental History of the Ruhr Basin, 1840–1990," *Environmental History Review* 18 (1994): 35–54.

3. Sheail, "Sustainable Management," 212.

4. For a discussion of economic, political, and environmental changes in Yorkshire, see Ray Hudson and David Sadler, "State Policies and the Changing Geography of the Coal Industry in the United Kingdom in the 1980s and 1990s," *Transactions of the Institute of British Geographers* 15 (1990): 435–54; Sheail, "'Never Again,'" 120–33; John Finch, "The Industrial Effluent Problem in Britain," *Water & Sewage Works* 104 (1957): 428–30. Regarding the Ruhr, see Stefan Goch, "Betterment without Airs: Social, Cultural, and Political Consequences of De-industrialization in the Ruhr," *International Review of Social History* 47, Supp. 10 (2002): 87–111; Brüggemeier, "A Nature Fit for Industry," 50–51; Franz-Josef Brüggemeier and Thomas Rommelspacher, *Blauer Himmel über der Ruhr: Geschichte der Umwelt im Ruhrgebiet, 1840–1990* (Essen: Klartext, 1992); and Klaus Georg Wey, *Umweltpolitik in Deutschland: Kurze Geschichte des Umweltschutz in Deutschland seit 1900* (Opladen: Westdeutscher Verlag, 1982), 173–81.

5. W. E. Tate and F. B. Singleton, *A History of Yorkshire, with Maps and Pictures* (Beaconsfield, England: Darwen Finlayson, 1967), 6–7, 47–65.

6. Dunbar quoted in Leopold Schua and Roma Schua, *Wasser: Lebenselement und Umwelt: Die Geschichte des Gewässerschutzes in ihrem Entwicklungsgang dargestellt und dokumentiert* (Munich: Verlag Karl Alber, 1981), 193–95, esp. 194. For a discussion of why industrial firms needed plentiful clean water, see Cioc, *The Rhine*, 88.

7. Quoted in Martin V. Melosi, *The Sanitary City: Urban Infrastructure from Colonial Times to the Present* (Baltimore: Johns Hopkins University Press, 2000), 51–56, esp. 54–55.

8. Ibid., 51–56.

9. Sheail, "Sustainable Management," 200–202.

10. Ibid., 201–3; and Melosi, *The Sanitary City*, 165.

11. Quoted in Sheail, "Public Interest and Self-interest," 149–70, esp. 151.

12. Ibid., 151; Elizabeth Porter, *Water Management in England and Wales* (Cambridge: Cambridge University Press, 1978), 116–18.

13. Sheail, "Sustainable Management," 197–215, esp. 201.

14. Ibid., 203–5.

15. Ibid.

16. Christopher Hamlin, "Muddling in Bumbledon: On the Enormity of Large Sanitary Improvements in Four British Towns, 1855–1885," *Victorian Studies* 32 (1988): 55–79.

17. Ibid., 73–75.

18. Quoted in ibid., 73.

19. I. W. Mendelsohn, "Water Supply and Purification Developments during 1934," *Water Works and Sewerage* 82 (1935): 2–6.

20. Porter, *Water Management in England and Wales*, 20–29, 116–18; Sheail, "Sustainable Management," 203–10; Sheail, "'Never Again,'" 117–33; Sheail, "Public Interest and Self-interest," 149–60.

21. Quoted in Sheail, "Public Interest and Self-interest," 159.

22. R. C. S. Walters, "Water Supply and Legislation in Britain," *Water & Sewage Works* 94 (1947): 336–37, esp. 336.

23. Sheail, "'Never Again,'" 122.

24. Sheail, "Sustainable Management," 205–8; Sheail, "Public Interest and Self-interest," 161–62.

25. Mendelsohn, "Water Supply and Purification Developments during 1934," 2–6.

26. Sheail, "'Never Again,'" 119; Porter, *Water Management in England and Wales*, 26–29, 118; Paul Smith, "Britain's Post-War Water Policy and Plans," *Water & Sewage Works* 96 (1949): 256–57, esp. 256. Smith's article discusses evidence that sanitary experts were involved in campaigns to reform the British political system. For more on that topic see Walters, "Water Supply and Legislation in Britain," 336.

27. Porter, *Water Management in England and Wales*, 26–29, 118; Sheail, "'Never Again,'" 122.

28. Porter, *Water Management in England and Wales*, 26–29, 118.

29. John Finch, "The Industrial Effluent Problem in Britain," *Water & Sewage Works* 3 (1957): 428–30; John Finch, "Report from Abroad: Problems of Atomic Energy Wastes and of 'Too Much' Industry," *Water & Sewage Works* 102 (1955): 86–88.

30. H. Fish, "River Pollution Prevention Succeeds in Britain," *Water & Sewage Works* 116 (1969): 336–40, esp. 338–39.

31. Finch, "The Industrial Effluent Problem in Britain," 428–30; John Finch, "Industrial Effluents in Municipal Sewers," *Water & Sewage Works* 105 (1958): 72–75.

32. Quoted in Sheail, "Public Interest and Self-interest," 167.

33. Sheail, "'Never Again,'" 117–34, esp. 131; Fish, "River Pollution Prevention," 336–40.

34. Regarding the British economy, see Walter Laqueur, *Europe in Our Time: A History, 1945–1992* (New York: Viking Press, 1992), 212–13. Regarding water consumption in the 1940s and 1950s, see W. G. V. Balchin, "A Water Use Survey," *Geographical Journal* 124 (1958): 476–82.

35. Finch, "The Industrial Effluent Problem in Britain," 428.

36. Fish, "River Pollution Prevention," 340.

37. John C. Rodda, "Physical Problems of the Urban Environment: A Symposium: Discussion," *Geographical Journal* 42.1 (1976): 74–75, esp. 75.

38. For various discussions of these changes, see Porter, *Water Management in England and Wales*, 118–30; Karen Bakker, "Privatizing Water, Producing Scarcity: The Yorkshire Drought of 1995," *Economic Geography* 76 (2000): 4–27; Fish, "River Pollution Prevention," 336–40.

39. John Ardill, "Managing Great Britain's Water," *Water & Sewage Works* 120 (1973): 38–40.

40. M. B. Beck, "Topic of Public Interest: Water Quality," *Journal of the Royal Statistical Society: Series A (General)* 147 (1983): 293–304, esp. 294; Rodda, "Physical Problems of the Urban Environment," 74–75.

41. Regarding the growth of environmental consciousness in Great Britain, see Ramachandra Guha, *Environmentalism: A Global History* (New York: Longman, 2000), 63–83; National Digital Archive of Datasets, *Environment Departments* (Richmond, England: National Archives, 2004), http://www.ndad.nationalarchives.gov.uk/AH/49/detail.html (accessed Apr. 2, 2007). See also Neil Carter and Philip Lowe, "Britain: Com-

ing to Terms with Sustainable Development?" 17–39; and Alan Butt Philip, "The European Union: Environmental Policy and the Prospects for Sustainable Development," 253–56, both in *Governance and Environment in Western Europe: Politics, Policy, and Administration*, ed. Kenneth Hanf and Alfe-Inge Jansen (New York: Longman, 1998).

42. Laqueur, *Europe in Our Time*, 212–15.

43. Ibid., 465–73; Bakker, "Privatizing Water," 4–27.

44. Bakker, "Privatizing Water," 20.

45. Ibid.; Karen Bakker, "Paying for Water: Water Pricing and Equity in England and Wales," *Transactions of the Institute of British Geographers* 126 (2001): 143–64.

46. Cioc, *The Rhine*, 21–23, 77–107; Norman J. G. Pounds, *The Ruhr: A Study in Historical and Economic Geography* (New York: Greenwood Press, 1968), 1–35, 53.

47. Cioc, *The Rhine*, 77–87; Pounds, *The Ruhr*, 96–114.

48. Wey, *Umweltpolitik in Deutschland*, 25; Pounds, *The Ruhr*, 120–27.

49. Quoted in Wey, *Umweltpolitik in Deutschland*, 31.

50. Ibid., 31–33; Peter Münch, *Stadthygiene im 19. und 20. Jahrhundert: Die Wasserversorgung, Abwasser, und Abfallbeseitigung unter besonderer Berücksichtigung Münchens* (Göttingen: Vandehoeck & Ruprecht, 1993), 57–61.

51. Regarding Great Britain, see Hamlin, "Muddling in Bumbledon," 55–79. Regarding Germany, see Raymond H. Dominick, *The Environmental Movement in Germany: Prophets and Pioneers, 1871–1971* (Bloomington: Indiana University Press, 1992), 12–13; Münch, *Stadthygiene*, 57–61.

52. D. A. Ramshorn, "Die Emschergenossenschaft," in *Fünfzig Jahre Emschergenossenschaft, 1906–1956*, ed. D. A. Ramshorn (Essen: Emschergenossenschaft, 1957), 38.

53. Cioc, *The Rhine*, 77–81.

54. Ibid., 77–101; Brüggemeier, "A Nature Fit for Industry," 36–50; Brüggemeier and Rommelspacher, *Blauer Himmel über der Ruhr*, 89–121.

55. Quoted in Brüggemeier and Rommelspacher, *Blauer Himmel über der Ruhr*, 89; Cioc, *The Rhine*, 77–92.

56. Brüggemeier and Rommelspacher, *Blauer Himmel über der Ruhr*, 89–100.

57. Ibid., 91–92.

58. Quoted in ibid., 92.

59. Cioc, *The Rhine*, 87–91; Brüggemeier and Rommelspacher, *Blauer Himmel über der Ruhr*, 89–99.

60. Brüggemeier and Rommelspacher, *Blauer Himmel über der Ruhr*, 96–97.

61. Wey, *Umweltpolitik in Deutschland*, 55–66.

62. For a full transcript of the Prussian Wastewater Law of 1913, see Wey, *Umweltpolitik in Deutschland*, 60; see also Wey's in-depth discussion of this law at 46–63.

63. Ramshorn, "Die Emschergenossenschaft," 36–42; Thomas Rommelspacher, "Das natürliche Recht auf Wasserverschmutzung: Geschichte des Wassers im 19. und 20. Jahrhundert," in *Besiegte Natur: Geschichte der Umwelt im 19. und 20. Jahrhundert*, ed. Franz-Josef Brüggemeier and Thomas Rommelspacher (Munich: Beck, 1989), 42–63.

64. Cioc, *The Rhine*, 94–95.

65. Ibid., 94–95.
66. Brüggemeier and Rommelspacher, *Blauer Himmel über der Ruhr*, 99.
67. Cioc, *The Rhine*, 95; Wey, *Umweltpolitik in Deutschland*, 78–79.
68. E. H. Wiegmann, "Die Abwasserreinigung im Emschergebiet," in *Fünfzig Jahre Emschergenossenschaft, 1906–1956*, ed. D. A. Ramshorn (Essen: Emschergenossenschaft, 1957), 240; Cioc, *The Rhine*, 92–100.
69. See full discussion in Wey, *Umweltpolitik in Deutschland*, 77–88, esp. 80.
70. Cioc, *The Rhine*, 92–93.
71. Ibid., 96–97; Brüggemeier and Rommelspacher, *Blauer Himmel über der Ruhr*, 101–2.
72. Brüggemeier, "A Nature Fit for Industry," 37–52, esp. 49.
73. Quoted in ibid., 50.
74. Dominick, *The Environmental Movement in Germany*, 140–43.
75. Ibid., 139–43; Wey, *Umweltpolitik in Deutschland*, 173–76; H. Rhode, "Wastewater Technology in Germany: Part I," *Water & Sewage Works* 106 (1959): 176–81.
76. Brüggemeier and Rommelspacher, *Blauer Himmel über der Ruhr*, 108–9; Brüggemeier, "A Nature Fit for Industry," 50–51; Pounds, *The Ruhr*, 227–61.
77. Wey, *Umweltpolitik in Deutschland*, 33–64, 173–81.
78. Brüggemeier, "A Nature Fit for Industry," 50–51.
79. Brüggemeier and Rommelspacher, *Blauer Himmel über der Ruhr*, 108–10.
80. Ibid., 108–24; Brüggemeier, "A Nature Fit for Industry," 37–54; Dominick, *The Environmental Movement in Germany*, 139–43, 187–93; Wey, *Umweltpolitik in Deutschland*, 180–82; Hendrik Seeger, "The History of German Waste Water Treatment," *European Water Management* 2 (1999): 51–56.
81. Seeger, "The History of German Waste Water Treatment," 51–56, esp. 54.
82. Federal Ministry for the Environment, Nature Conservation and Nuclear Safety (Bundesministerium für Umwelt, Naturschutz und Reaktorsicherheit [BMU]), *Waste Water Law: Federal Water Act, Waste Water Charges Act, Waste Water Ordinance* (2003), http://www.bmu.de/files/pdfs/allgemein/application/pdf/wastewater.pdf (accessed Apr. 2, 2007); Klaus Seifert and T. Clark Lyons, "Water Quality Law in the Federal Republic of Germany," *Journal of the Water Resources Planning and Management Division* 102.1 (1976): 23–33; K. R. Imhoff, "Water Pollution Control Measures and Water Quality Development in the Ruhr Catchment, 1972–1992," *Water, Science and Technology* 2.5–6 (1995): 209–16; and Heinrich Pehle and Alfe-Inge Jansen, "Germany: The Engine in European Environmental Policy?" in *Governance and Environment in Western Europe: Politics, Policy, and Administration*, ed. Kenneth Hanf and Alfe-Inge Jansen (New York: Longman, 1998), 86–99.
83. Imhoff, "Water Pollution Control Measures," 209–12; Brüggemeier, "A Nature Fit for Industry," 35–54; Harro Bode, *River Basin Management, Legal Background, Quality Monitoring, Purification Techniques and Costs* (n.d.), http://www.unesco.org.uy/phi/libros/VIJornadas/C38.pdf, 1–14 (accessed Apr. 3, 2007).
84. Imhoff, "Water Pollution Control Measures," 211–13.

85. Hudson and Sadler, "State Policies and the Changing Geography of the Coal Industry," 435–54. See also "A Small City in Germany Mirrors a National Malaise," *New York Times,* Oct. 10, 2004, http://www.lexisnexis.com.dax.lib.unf.edu/us/lnacademic/ auth/checkbrowser.do (accessed May 28, 2007).

86. Brüggemeier, "A Nature Fit for Industry," 35–53; Bode, *River Basin Management,* 11–12.

87. Sheail, "Sustainable Management," 212.

88. Carter and Lowe, "Britain: Coming to Terms," 17–35.

Chapter 8: Saving the Rhine

1. International Commission for the Protection of the Rhine against Pollution, *Ecological Master Plan for the Rhine/"Salmon 2000"* (Koblenz: ICPR Technical-Scientific Secretary, 1990), 5 (cited hereafter as Rhine Commission, *Salmon 2000*); Marco Verweij, *Transboundary Environmental Problems and Cultural Theory: The Protection of the Rhine and the Great Lakes* (New York: Palgrave, 2000), 71; Mark Cioc, *The Rhine: An Eco-Biography, 1815–2000* (Seattle: University of Washington Press, 2002), 3.

2. Cioc, *The Rhine,* 22–23; T. H. Elkins and P. K. Marstrand, "Pollution of the Rhine and Its Tributaries," in *Regional Management of the Rhine: Papers of a Chatham House Study Group* (London: Chatham House, 1975), 50–69; Menno T. Kamminga, "Who Can Clean Up the Rhine: The European Community or the International Rhine Commission?" in *The Legal Regime of International Rivers and Lakes,* ed. Ralph Zacklin et al. (The Hague: Martinus Nijhoff, 1981), 371–87; Beate Kretteck, "Der Rhein—Kloake, Giftkanal oder Ökosystem: Dokumentation zur ökologischen Situation," in *Der Rhein—Mythos und Realität eines europäischen Stromes,* ed. Hans Boldt (Cologne: Rheinland Verlag, 1988), 131–38.

3. Cited in Verweij, *Transboundary Environmental Problems,* 72.

4. Verweij, *Transboundary Environmental Problems,* 95; Cioc, *The Rhine,* 182–98; International Commission for the Protection of the Rhine against Pollution, *Rhine Action Programme* (Koblenz: Internationale Kommission zum Schutze des Rheins gegen Verunreinigung, 1987).

5. Recent scholarship has called into question the anti-modernism and agrarian-romanticism of pre–World War II German nature conservation and landscape preservation, emphasizing instead its search for an alternative, sustainable environmental modernity. For a survey of these issues, see Thomas Lekan and Thomas Zeller, "Introduction: The Landscape of German Environmental History," in *Germany's Nature: Culture Landscapes and Environmental History,* ed. Thomas Lekan and Thomas Zeller (New Brunswick, NJ: Rutgers University Press, 2005), 1–16.

6. Cioc, *The Rhine,* 177; Axel Goodbody, "Anxieties, Visions, and Realities: Environmentalism in Germany," in *The Culture of German Environmentalism: Anxieties, Visions, Realities,* ed. Axel Goodbody (New York: Berghahn, 2002), 33–34; Sandra Chaney, "Visions and Revisions of Nature: From the Protection of Nature to the Invention of the

Environment in the Federal Republic of Germany, 1945–1975" (Ph.D. diss., University of North Carolina–Chapel Hill, 1996), 22.

7. Chaney, "Visions and Revisions of Nature," 3; Raymond Dominick, *The Environmental Movement in Germany: Prophets and Pioneers, 1871–1971* (Bloomington: Indiana University Press, 1992), 182–214; Monika Bergmeier, *Umweltgeschichte der Boomjahre, 1949–1973: Das Beispiel Bayern* (Münster: Waxmann, 2002), 9–22, 256–57. On the fifties syndrome, see Christian Pfister, *Das 50er Syndrom: Der Weg in die Konsumgesellschaft* (Bern: Paul Haupt, 1995).

8. Recent works that detail the aesthetic, nationalist, and cultural foundations of prewar German nature conservation include Werner Hartung, *Konservative Zivilisationskritik und regionale Identität—Am Beispiel der niedersächsischen Heimatbewegung 1895 bis 1919* (Hannover: Hahn'sche Buchhandlung, 1991); William Rollins, *A Greener Vision of Home: Cultural Politics and Environmental Reform in the German Heimatschutz Movement* (Ann Arbor: University of Michigan Press, 1997); Andreas Knaut, *Zurück zur Natur! Die Wurzeln der Ökologiebewegung* (Greven: Kilda-Verlag, 1993); and Thomas Lekan, *Imagining the Nation in Nature: Landscape Preservation and German Identity* (Cambridge, MA: Harvard University Press, 2004).

9. For a good summary of *Kulturkritik* applicable to environmental history, see Thomas Rohkrämer, *Eine andere Moderne? Zivilisationskritik, Natur und Technik in Deutschland, 1880–1933* (Paderborn: Schöningh, 1999).

10. See Paul R. Josephson, *Industrialized Nature: Brute Force Technology and the Transformation of the Natural World* (Washington, DC: Island Press, 2002).

11. Peter Thorsheim, "Interpreting the London Fog Disaster of 1952," in *Smoke and Mirrors: The Politics and Culture of Air Pollution*, ed. E. Melanie DuPuis (New York: New York University Press, 2004), 154–69, esp. 154. DuPuis's introduction to this anthology offers a thoughtful summary of the constructivist position on air pollution issues past and present.

12. Mary Douglas, *Purity and Danger: An Analysis of Concepts of Pollution and Taboo* (Washington, DC: Praeger, 1966), 1–3. On the cultural significance of water, see also Susan Anderson, "Introduction: The Rhetorical Power of Water," in *Water, Culture, and Politics in Germany and the American West*, ed. Susan Anderson and Bruce Tabb (New York: Peter Lang, 2001), 1–10.

13. On the contradictory cultural tendencies during this period, see the provocative essays in Hanna Schissler, ed., *The Miracle Years: A Cultural History of West Germany, 1949–1968* (Princeton, NJ: Princeton University Press, 2001).

14. Memorandum of Oberpräsident of the North Rhine Province concerning Rhine Pollution, Aug. 15, 1946, fol. 7; Memorandum of Headquarters of Military Government, Land North Rhine–Westphalia to Baurat Strümpfel in the Düsseldorf Landeshaus, Aug. 24, 1946, fol. 9, both in Hauptstaatsarchiv Nordrhein-Westphalen (hereafter NRW), nos. 268–518.

15. On the foundations of the Rhine Commission, see Cioc, *The Rhine*, 178–79; Otto Jaag, "Internationale Zusammenarbeit im europäischen Gewässerschutz," in *Fachblatt der Gastechnik und Gaswirtschaft sowie für Wasser und Abwasser* 44.100 (Oct. 1959): 1135–36;

Fritz Lippert, "Die Reinhaltung des Rheins—Eine internationale Aufgabe" (Dec. 1959) and "Entstehungsgeschichte der International Kommission zum Schutze des Rheins gegen Verunreinigung" (1959), both in NRW, nos. 354–672; Kenneth J. Langran, "International Water Quality Management: The Rhine River as a Study in Transfrontier Pollution Control" (Ph.D. diss., University of Wisconsin–Madison, 1979), 126–27; Torsten Mick and Michael Tretter, "Der Rhein und Europa," in *Der Rhein: Mythos und Realität eines europäischen Stromes*, ed. Hans Boldt (Cologne: Rheinland, 1988), 42–43. The European Community joined the Rhine Commission as an institutional member in 1979.

16. "Die Verunreinigung des Rheins in Nordrhein-Westfalen: Eine Denkschrift aus dem Lande Nordrhein-Westfalen" (1956), NRW, nos. 354–406.

17. From "Cologne," quoted in Verweij, *Transboundary Environmental Problems*, 109.

18. See Chaney, "Visions and Revisions of Nature," 58–60; Robert Evans, *Death in Hamburg: Society and Politics in the Cholera Years, 1830–1910* (Oxford: Clarendon, 1987); Thomas Kluge and Engelbert Schramm, *Wassernöte: Umwelt- und Sozialgeschichte des Trinkwassers* (Aachen: Alano, 1986), 9–38; Werner Reh, "Die ökologische Folgen der Industrialisierung am Rhein," in *Der Rhein: Mythos und Realität eines europäischen Stromes*, ed. Hans Boldt (Cologne: Rheinland, 1988), 120–21.

19. Cioc, *The Rhine*, 139–40; Reh, "Ökologische Folgen," 121. This pattern of poor investment in sewage treatment was a national problem; in the 1950s less than half the residents of the FRG were attached to sewage treatment of any kind. See Charles E. Closmann, "Modernizing the Waters: Water Pollution and the Harbor Economy in Hamburg, Germany, 1900–1961" (Ph.D. diss., University of Houston, 2001), 215, 239–41.

20. Cioc, *The Rhine*, 109–43.

21. Chaney, "Visions and Revisions of Nature," 59.

22. Kretteck, "Der Rhein—Kloake, Giftkanal oder Ökosystem?" 136; Cioc, *The Rhine*, 94–97.

23. Cioc, *The Rhine*, 89–97.

24. Cioc, *The Rhine*, 3, 11–12, 47–75; David Blackbourn, *The Conquest of Nature: Water, Landscape, and the Making of Modern Germany* (New York: Norton, 2006), 93–119.

25. Reinhard Demoll, *Heimat*, 2 vols. (Basel: Faunus, 1958–59), 2:457.

26. Cioc, *The Rhine*, 126–27. On nineteenth-century water pollution more broadly, see Thomas Rommelspacher, "Das Natürliche Recht auf Wasserverschmutzung," in *Besiegte Natur: Geschichte der Umwelt im 19. und 20. Jahrhundert*, ed. Franz-Josef Brüggemeier and Thomas Rommelspacher, 2nd ed. (Munich: C. H. Beck, 1989), 62–63; Jürgen Büchenfeld, *Flüsse und Kloaken: Umweltfragen im Zeitalter der Industrialisierung, 1870–1918* (Stuttgart: Klett-Cotta, 1997); and Beate Olmer, *Wasser. Historisch: Zu Bedeutung und Belastung des Umweltmediums im Ruhrgebiet, 1870–1930* (Frankfurt: Peter Lang, 1998).

27. Cioc, *The Rhine*, 126–41; "Die Verunreinigung des Rheins in Nordrhein-Westfalen," nos. 354–406.

28. Dominick, *The Environmental Movement in Germany*, 141. See also Bundesministerium für Verkehr, "Die Ölverschmutzung des Rheins und Vorschläge zu ihrer Bekämpfung auf dem Gebiet der Binnenschifffahrt" (Bonn, 1958), NRW, nos. 268–530.

29. Cioc, *The Rhine*, 12; Reh, "Ökologische Folgen," 118.

30. Kretteck, "Der Rhein—Kloake, Giftkanal oder Ökosystem," 136–38. See also numerous documents on oil pollution, radioactive contamination, and other issues in NRW, nos. 132–368, 799, 803, and 890.

31. Colin Riordan, "Green Ideas in Germany: A Historical Survey," in *Green Thought in German Culture: Historical and Contemporary Perspectives*, ed. Colin Riordan (Cardiff: University of Wales Press, 1997), 27–28.

32. Dominick, *The Environmental Movement in Germany*, 141–42.

33. Ibid., 187. On the "risk society" resulting from fears of unseen radioactivity and chemical pollution, see also Ulrich Beck, *Risikogesellschaft: Auf dem Weg in eine andere Moderne* (Frankfurt: Suhrkamp, 1986).

34. Dominick, *The Environmental Movement in Germany*, 165.

35. "Tod im Strom," *Der Spiegel* 24.9 (Feb. 23, 1970): 46.

36. Cioc, *The Rhine*, 135–37.

37. "Reinhaltung des Rheins," *Fachblatt für Gastechnik und Gaswirtschaft sowie für Wasser und Abwasser* 44.100 (Oct. 1959): 1130, my translation. Unless otherwise noted, translations from German to English throughout this essay are my own.

38. Quoted in Dominick, *Environmental Movement in Germany*, 143.

39. On interwar holism in German biology, see Anne Harrington, *Reenchanted Science: Holism in German Culture from Wilhelm II to Hitler* (Princeton: Princeton University Press, 1996). On the relationship between nature conservation and the discipline of ecology, see Thomas Potthast, "Wissenschaftliche Ökologische und Naturschutz: Szene einer Annäherung," in *Naturschutz und Nationalsozialismus*, ed. Joachim Radkau and Frank Uekötter (Frankfurt: Campus, 2003), 225–56.

40. Cioc, *The Rhine*, 92.

41. August Thienemann, *Der Mensch als Glied und als Gestalter der Natur* (Jena and Leipzig: Wilhelm Gronau, 1944), 8.

42. August Thienemann, *Leben und Umwelt: Vom Gesamthaushalt der Natur* (Hamburg: Rohwohlt, 1956).

43. August Thienemann, "Kein Hahn darf tropfen: Das Dilemma der Wasserwirtschaft," *Deutsche Zeitung*, Nov. 3, 1956, quoted in Dominick, *The Environmental Movement in Germany*, 137.

44. August Thienemann, *Erinnerungen und Tagebuchblätter eines Biologen: Ein Leben im Dienste der Limnologie* (Stuttgart: E. Schweizerbar, 1959), 395–96.

45. Dominick, *The Environmental Movement in Germany*, 140.

46. Siegfried Balke, quoted in Dominick, *The Environmental Movement in Germany*, 140–41.

47. See Dominick, *The Environmental Movemement in Germany*, 142–43.

48. Ibid., 143.

49. Ibid., 139–40.

50. Ibid., 187–88; "Wasser: Alarm in der Leitung," *Der Spiegel* 13.47 (Nov. 18, 1959): 36–47.

51. "Wasser: Alarm in der Leitung," 37–38.

52. Ibid., 46–47; "Schlag ins Wasser," *Der Spiegel* 13.49 (Dec. 2, 1959): 3.

53. "Der Rheinstrom ist gefährlich krank," *Düsseldorfer Nachrichten* 73.98 (1952).

54. Verweij, *Transboundary Environmental Problems*, 140.

55. "Fischkatastrophe in der Ahr," *Westfälische Nachrichten*, no. 108 (May 10, 1957). On public awareness of fish kills, see also letters to the editor of *Der Spiegel*, in "Schlag ins Wasser," 3–5.

56. See Kai F. Hünemörder, *Frühgeschichte der globalen Umweltkrise und die Formierung der deutschen Umweltpolitik (1950–1973)* (Wiesbaden: Franz Steiner Verlag), 86.

57. "Der Rheinstrom ist gefährlich krank," *Düsseldorfer Nachrichten* 73.98 (1952). On the massive Rhine fish kill of June 1969, see Hünemörder, *Die Frühgeschichte der globalen Umweltkrise*, 84–87.

58. Rhine Commission, "Salmon 2000," 3; Reh, "Ökologische Folgen," 117–18; Blackbourn, *The Conquest of Nature*, 105–7.

59. Dominick, *The Environmental Movement in Germany*, 158–69; Chaney, "Visions and Revisions of Nature," 261.

60. Dominick, *The Environmental Movement in Germany*, 158–69.

61. Chaney, "Visions and Revisions of Nature"; Erich Hornsmann, *Sonst Untergang: Die Antwort der Erde auf die Missachtung ihrer Gesetze* (Rheinhausen: Verlaganstalt Rheinhausen, 1951), 347–80. See also Hünemörder, *Die Frühgeschichte der globalen Umweltkrise*, 16, 34–47.

62. Chaney, "Visions and Revisions of Nature," 294–95, 358–70.

63. On Riehl and his impact on German nature conservation and homeland protection movements, see Rollins, *A Greener Vision of Home*; Celia Applegate, *A Nation of Provincials: The German Idea of Heimat* (Berkeley: University of California Press, 1990), 34–41; and Jasper von Altenbockum, *Wilhelm Heinrich Riehl 1823–1897: Sozialwissenschaft zwischen Kulturgeschichte und Ethnographie* (Cologne: Böhlau, 1994).

64. On the nationalization of the Rhine, see Lekan, *Imagining the Nation in Nature*, 19–73; Hans Boldt, "Deutschlands hochschlagende Pulsader," in *Der Rhein: Mythos und Realität eines europäischen Stromes*, ed. Hans Boldt (Cologne: Rheinland, 1988), 30–32.

65. The literature on the concept of *Heimat* in German culture is enormous and cannot be cited here in its entirety. Works detailing its history and cultural significance include Applegate, *A Nation of Provincials;* Ina-Marie Greverus, *Auf der Suche Nach Heimat* (Munich: C. H. Beck, 1979); and Jost Hermand and James Steakley, eds., *Heimat, Nation, Fatherland: The German Sense of Belonging* (New York: Peter Lang, 1996).

66. See Lekan, *Imagining the Nation in Nature*, 121–41.

67. On the discrepancy between Nazi conservationist rhetoric and their prioritization of environmentally destructive rearmament, land reclamation, and road building, see Joachim Radkau and Frank Uekötter, eds., *Naturschutz und Nationalsozialismus* (Frankfurt: Campus, 2003); Lekan, *Imagining the Nation in Nature*, 204–12; Mark Cioc, Franz-Josef Brüggemeier, and Thomas Zeller, eds., *How Green Were the Nazis: Nature, Environment, and Nation in the Third Reich* (Athens: Ohio University Press, 2005); and Frank Uekötter, *The Green and the Brown: A History of Nature Conservation in Nazi Germany* (Cambridge: Cambridge University Press, 2006).

68. On the landscape protection plans, see Lekan, *Imagining the Nation in Nature*,

184–89; "Endgültiger Entwurf, Verordnung zum Schutz der Landschaft des Mittelrheins" (1941?), in Archiv des Landschaftverbandes Rheinland (hereafter ALVR), no. 11241.

69. Franz Wolff Metternich, Hanna Adenauer, and Josef Ruland, eds., *Festschrift für Franz Graf Wolff Metternich* (Neuss: Verlag Gesellschaft für Buchdruckerei AG, 1973), 55–58; "Endgültiger Entwurf" (1941?) and "Begründung einer für das Rheintal zu erlassenden Schutzverordnung nebst kritischer Würdigung des Entwurfes Becker und der Abänderungsanträge der beteiligten Behörden," 41, ALVR, no. 11241. See also Walther Schoenichen, *Naturschutz als völkische und internationale Kulturaufgabe* (Jena: Gustav Fischer, 1942), 33, 55.

70. "Endgültiger Entwurf" (1941?).

71. Jens Ivo Engels, "'Hohe Zeit' und 'dicker Strich': Vergangenheitsdeutung und bewahrung im westdeutschen Naturschutz nach dem Zweiten Weltkrieg," 262–404; and Stefan Körner, "Kontinuum und Bruch: Die Transformation des naturschützerischen Aufgabenverständnisses nach dem Zweiten Weltkrieg," 405–34, both in *Naturschutz und Nationalsozialismus*, ed. Joachim Radkau and Frank Uekötter (Frankfurt: Campus, 2003).

72. Dominick, *The Environmental Movement in Germany*, 197. See memorandum from Dr. Siegmond to the North Rhine–Westphalian Cultural Ministry concerning Lehrfilm "Natur in Gefahr," June 5, 1957, fols. 25–27; and "Natur in Gefahr: Der Kulturfilm von dem man spricht—Ein Film der alle angeht," fol. 175, both in NRW, nos. 260–22.

73. Kulturfilmstunde für Volksschule und höhere Schulen, NRW, nos. 260–22, fol. 178.

74. Demoll, *Heimat*, 400.

75. On these continuities, see Thomas Rohkrämer, "Contemporary Environmentalism and Its Links with the German Past," in *The Culture of German Environmentalism: Anxieties, Visions, Realities*, ed. Axel Goodbody (New York: Berghahn, 2002), 47–62.

76. On the reorientation of bourgeois culture and German conservatism after the war, see Anson Rabinbach, "Restoring the German Spirit: Humanism and Guilt in Post-War Germany," in *German Ideologies since 1945: Studies in the Political Thought and Culture of the Bonn Republic*, ed. Jan-Werner Müller (New York: Palgrave Macmillan, 2003), 23–39.

77. Demoll, *Heimat*, 449.

78. Such reconciliation with modernization and increasingly technocratic tendencies were common to German conservatism after 1945. See, for example, Dirk von Laak, "From the Conservative Revolution to Technocratic Conservatism," in *German Ideologies since 1945: Studies in the Political Thought and Culture of the Bonn Republic*, ed. Jan-Werner Müller (New York: Palgrave Macmillan, 2003), 147–60; and Hünemörder, *Die Frühgeschichte der globalen Umweltkrise*, 88–89.

79. Ministerialrat Piperek, "Mensch und Natur," *Unser Wasser, VDG Mitteilungen*, no. 4–5 (1956): 1; see also Jürgen Schwoebel, "Die biologische Gliederung des Rheinstromes," *GWF* 44.100 (Oct. 30, 1959): 1130–35.

80. Report of the North Rhine–Westphalian Agriculture and Forestry Ministry, "Bekämpfung der Gewässerverschmutzung," Feb. 15, 1957, fol. 3, NRW nos. 354–407.

81. Horst Klosterkemper, "Die Reinhaltung der Bundeswasserstrassen, insbesondere des Rheins aus der Sicht Nordrhein-Westfalens," *Wasser und Boden* 12.6 (1960): 1–4, esp. 1.

82. Douglas, *Purity and Danger*, 3–4. German engineers' interest in promoting a balance between technology and nature, *Zivilisation* and *Kultur*, can also be traced back to the prewar period. See Hans-Liudger Dienel, *Herrschaft über die Natur? Naturvorstellungen deutscher Ingenieure, 1871–1914* (Stuttgart: Verlag für Geschichte der Naturwissenschaften und der Technik, 1992).

83. Mick and Tretter, "Der Rhein und Europa," 35. See also Ernst Moritz Arndt, *Der Rhein, Deutschlands strom, aber nicht Deutschlands gränze* (Leipzig, 1813).

84. Mick and Tretter, "Der Rhein und Europa," 41–42; Cioc, *The Rhine*, 177.

85. Quoted in Mick and Tretter, "Der Rhein und Europa," 38.

86. Ernst Noth, *Bridges over the Rhine* (New York: Henry Holt, 1947).

87. Mick and Tretter, "Der Rhein und Europa," 42.

88. Cioc, *The Rhine*, 178.

89. Otto Jaag, "Internationale Zusammenarbeit im europäischen Gewässerschutz," *GWF* 100.44 (Oct. 30, 1959): 1135–37, esp. 1135.

90. Ibid., 1136; see also Internationale Kommission zum Schutze des Rheins gegen Verunreinigung, "Entwurf für die Geschäftsordnung der Kommission" (Aug. 1959) in NRW, nos. 354–691.

91. Föderation Europäischer Gewässerschutz, *Unser Wasser*, no. 10–11 (1956): 1.

92. Verweij, *Transboundary Environmental Problems*, 72.

93. Ibid., 83–90; Cioc, *The Rhine*, 178–80.

94. Verweij, *Transboundary Environmental Problems*, 85; Reh, "Ökologische Folgen," 123. On the debates about the water law, see Deutscher Bundestag 3. Wahlperiode, Drucksache 46, Entwurf eines Gesetzes zur Reinhaltung der Bundeswasserstrassen (WStrRG) (Nov. 30, 1957), in NRW, nos. 268–522; "Milliarden für den Rhein: Ein Sanierungsplan gegen die Verschmutzung," NRW, nos. 132–800.

95. Verweij, *Transboundary Environmental Problems*, 91–100.

96. Rhine Commission, *Salmon 2000*; Verweij, *Transboundary Environmental Problems*, 94–95.

97. Cioc, *The Rhine*, 185–98.

98. Chaney, "Visions and Revisions of Nature," 358–444; Bergmeier, *Umweltgeschichte der Boomjahre*, 255; Hünemörder, *Die Frühgeschichte der globalen Umweltkrise*, 189–90.

99. Mick and Tretter, "Der Rhein und Europa," 35.

100. Chaney, "Visions and Revisions of Nature," 437–44; Riordan, "Green Ideas in Germany," 59–61.

101. On this point, see the essays in Frank Zelko, ed., *From Heimat to Umwelt: New Perspectives on German Environmental History*, supplement 3 of the *Bulletin of the German Historical Institute* (2006).

102. On the history of the Greens, see Thomas Poguntke, *Alternative Politics: The German Green Party* (Edinburgh: Edinburgh University Press, 1993); and Riordan, ed., *Green Thought in German Culture*.

103. Rohkrämer, "Contemporary Environmentalism and Its Links with the German Past," 59–61.

Chapter 9: Postwar Perceptions of German Rivers

1. The term *pre-Alpine rivers* refers to streams and brooks situated in the Alpine foothills, a sometimes level, sometimes mountainous region in the vicinity of the Alps, which is characterized by high altitude. Most pre-Alpine rivers have their sources in the Alps themselves.

2. Günther Garbrecht, *Wasser: Vorrat, Bedarf und Nutzung in Geschichte und Gegenwart* (Reinbek: Rowohlt, 1985), 197; Bernhard Stier, *Staat und Strom: Die politische Steuerung des Elektrizitätssystems in Deutschland 1890–1950* (Mannheim: Verlag Regionalkultur, 1999), 16; Wilhelm Füßl, *Oskar von Miller 1854–1934: Eine Biographie* (Munich: C. H. Beck, 2005), 46–140.

3. Königliche Oberste Baubehörde, ed., *Die Wasserkräfte Bayerns: Text* (Munich: Piloty and Loehle, 1907), 11.

4. The total inclination of the Lech, for example, is 1,460 meters, of which 404 meters belong to the state of Bavaria. Water supply fluctuates strongly, however, with an annual mean of 64.2 m^2/s at Füssen and 124.6 m^2/s at Rain (Staatsministerium des Innern, Oberste Baubehörde, Abteilung für Wasserkraftnutzung und Elektrizitätsversorgung, ed., *Wasserkraftausnützung in Bayern* [Munich: Wolf, 1926], 31–32).

5. Oberste Baubehörde, *Wasserkraftausnützung in Bayern*, i–xvi. Reservoirs not only balance low water tables in winter but also help to control floodwaters in the spring.

6. In 1946 only 30 percent of the amount of coal imported ten years earlier was available (Bayerisches Hauptstaatsarchiv [hereafter BHStA], Bayerische Staatskanzlei [hereafter StK] 14650: Wolf to Fischer, June 14, 1947).

7. BHStA, Bayerische Landesstelle für Naturschutz (hereafter LfN) 37: circular Fischer, Jan. 22, 1949; Fischer to Kraus, Mar. 31, 1949; Kraus to Ehard, Feb. 17, 1949; StK 14653: Bayernwerk AG, *Memorandum zur Frage der zukünftigen Gestaltung der bayerischen Stromversorgung*, Feb. 19, 1949; "Die Situation der bayerischen Stromversorgung," *Bayerischer Staatsanzeiger*, Dec. 3, 1949, 1; Stephan Deutinger, "Eine 'Lebensfrage für die bayerische Industrie': Energiepolitik und regionale Energieversorgung 1945 bis 1980," in *Bayern im Bund 1*, ed. Thomas Schlemmer and Hans Woller (Munich: Oldenbourg, 2001), 37–41.

8. BHStA, LfN 45: Huber to Bayerisches Staatsministerium des Innern (hereafter MInn), May 12, 1948; LfN 46: Kraus to MInn, Nov. 11, 1954.

9. BHStA, LfN 37: LfN to MInn, Nov. 28, 1949; Erwin Hofmann, "Gutachten über das geplante Partnach-Kraftwerk Werdenfels," in *Die Partnachklamm ist in Gefahr!* ed. DAV-Sektion Garmisch-Partenkirchen (Garmisch-Partenkirchen: Alpenverein Sektion Garmisch-Partenkirchen, 1949), 12–15; Otto Kraus, "Naturschutz und Wasserkraftnutzung in Bayern" (1952), in *Zerstörung der Natur: Unser Schicksal von morgen: Der Naturschutz in dem Streit der Interessen: Ausgewählte Abhandlungen und Vorträge* (Nuremberg: Glock und Lutz, 1966), 167–69. Unless otherwise noted, all essays by Otto Kraus are quoted from this 1966 collection of his work; the year in which each essay was originally published is given in parentheses after its title. Otto Kraus (1905–1984) was commissioner of the Bayerische Landesstelle für Naturschutz, a nature conservation

agency that acted in an advisory capacity to the state authorities. In the story of the Lech, he was the most important actor on the conservationists' side.

10. "Bayerns Kraftwerke decken den steigenden Strombedarf," *Süddeutsche Zeitung*, Nov. 17, 1954, 7; BHStA, StK 14655: Ministerialdirigent Krauss: *Kurzbericht über die Bedeutung der Pumpspeicheranlagen im Rahmen der öffentlichen Elektrizitätsversorgung Bayerns*, June 29, 1955; Bayerisches Staatsministerium für Wirtschaft und Verkehr, ed., *Stand und Entwicklung der bayerischen Energiewirtschaft* (Munich: Bayer. Staatsministerium f. Wirtschaft u. Verkehr, 1964).

11. BHStA, LfN 37: Otto Kraus to editors of newspaper *Münchner Merkur*, Dec. 17, 1949; "Verschiedene Mitteilungen," *Blätter für Naturschutz* 30 (1950): 23; Otto Kraus, "Energiewirtschaft der Alpenländer im Umbruch—Wasserkraftnutzung am Ende?" (1954), in *Zerstörung der Natur*, 170–78; Otto Kraus, *Bis zum letzten Wildwasser: Gedanken über Wasserkraftnutzung und Naturschutz im Atomzeitalter* (Aachen: Georgi, 1960), 24–27.

12. For a full description of this conflict, as well as a detailed analysis of the courses of action of the protagonists, see Ute Hasenöhrl, "Conflicts between Economic and Conservation Interests Concerning Hydroelectric Power Plants: The Case of the River Lech (Germany)," in *Proceedings of the Third International Conference of the IWHA, Alexandria (Egypt), 11–14 December 2003*, ed. International Water History Association (Bergen: International Water History Association, 2003). This publication is available only on CD-ROM.

13. I do not support Monika Bergmeier's argument that conservationists limited themselves to traditional arguments centering on nature conservation and *Heimatschutz* (homeland protection); see Monika Bergmeier, *Umweltgeschichte der Boomjahre 1949–1973: Das Beispiel Bayern* (Münster: Waxmann, 2002), 158.

14. E.g., BHStA, StK 13773: Verein zum Schutze der Alpenpflanzen und -Tiere to Hoegner, June 1, 1955; Seifert to Feneberg, Sept. 5, 1955; LfN 46: Micheler to mayor of Burggen, Nov. 27, 1959; "Alarm bei den Naturfreunden," *Landsberger Nachrichten*, Dec. 1, 1955; "Der Landschaftsmord muß ein Ende haben," *Schwäbische Landeszeitung*, May 15, 1959, 19.

15. BHStA, LfN 45: Kraus to Regierungsstelle für Naturschutz in Oberbayern, June 17, 1952; Anton Micheler, "Der Lech: Bild und Wandel einer voralpinen Flusslandschaft," *Jahrbuch des Vereins zum Schutze der Alpenpflanzen und -Tiere* (1953): 53–68.

16. Otto Kraus, "Stau-Kraftwerke retten kranke Flusslandschaften" (1957), in *Zerstörung der Natur*, 129–34.

17. BHStA, LfN 45: Kraus to Regierungsstelle für Naturschutz in Oberbayern, Aug. 23, 1952; Sepp to Regierung von Oberbayern, Jan. 28, 1953; LfN 46: Kraus to MInn, Apr. 14, 1954; Kraus to MInn, Dec. 15, 1960; Kraus to MInn, Mar. 1, 1962; "Was geht am oberen Lech vor?" *Blätter für Naturschutz* 39.1–2 (1959): 29; "Noch einmal: Was geht am oberen Lech vor?" *Blätter für Naturschutz* 39.3–4 (1959): 59–60.

18. BHStA, LfN 46: Kraus to MInn, Jan. 19, 1960; von Bomhard to Ehard, July 27, 1960; Landrat Hilger: "Dr. Hilger: Kreistagspolitik und Millionen-'Spende,'" *Schongauer Nachrichten*, July 4, 1962, 5–6; StK 13773: Seifert to Hoegner, May 17, 1955; StK 17021:

Kraus to MInn, June 24, 1963; "Wir kämpfen um den oberen Lech!" *Süddeutsche Zeitung* and *Münchner Merkur,* January 18, 1960; "Noch einmal," 59–60.

19. Nuclear power was not regarded as potentially risky; see BHStA, StK 13773: Verein zum Schutze der Alpenpflanzen und -Tiere to Hoegner, June 1, 1955; LfN 46: Kraus to Regierungsstelle für Naturschutz in Oberbayern, Feb. 5, 1962; Kraus, *Bis zum letzten Wildwasser,* 24–31; "Oberer Lech und Energiewirtschaft," *Blätter für Naturschutz* 42.3 (1963): 63–64.

20. Staatsarchiv München (hereafter StAM), Landratsamt Landsberg 193520: Fischer to BAWAG, Mar. 23, 1950; BHStA, LfN 46: Deinlein to Kraus, Jan. 21, 1960.

21. Bayerische Wasserkraftwerke AG, ed., *50 Jahre BAWAG 1940–1990: Natur und Energie in Harmonie* (Munich: Bayerische Wasserkraftwerke AG, 1990), 12; Bergmeier, *Umweltgeschichte der Boomjahre,* 157.

22. In contrast to Uekötter, however, I would not go so far as to say that tourism was used merely as a strategic argument; compare Frank Uekötter, *Naturschutz im Aufbruch: Eine Geschichte des Naturschutzes in Nordrhein-Westfalen 1945–80* (Frankfurt: Campus, 2004).

23. BHStA, StK 13775: *Wächter: Die Einleitung des Rissbaches in den Walchensee,* May 30, 1947; Ludwig A. Haimerl, *Speicherkraftwerk Rosshaupten* (Munich: Bayerische Wasserkraftwerke AG, 1961), 20; Josef Harlander, "Umweltschutz durch Kraftwerke in der Salzach," *Alt-Neuöttinger Anzeiger,* Sept. 3, 1977.

24. "It is more important to preserve this pristine nature for the whole German people than to produce the tiny amount of ten kilowatt hours per capita of the Bavarian population, which could also be achieved differently" (StAM, Landratsamt Garmisch-Partenkirchen 199586: Bund Naturschutz to Bayerischer Landtag, May 19, 1947). See also BHStA, LfN 64: Spindler, "Leserbrief," *Hochland-Bote,* June 8, 1949; "Alarm bei den Naturfreunden," *Landsberger Nachrichten,* Dec. 1, 1955; BHStA, LfN 63: Lense to Hoegner, Nov. 3, 1959; Kraus, *Bis zum letzten Wildwasser,* 18–21. Here and elsewhere in this essay, all English translations from the German are my own unless otherwise noted.

25. Bundesministerium für Umwelt, Naturschutz und Reaktorsicherheit, ed., *Umweltpolitik: Wasserwirtschaft in Deutschland: Teil 1: Grundlagen* (Bonn: Bundesministerium für Umwelt, Naturschutz und Reaktorsicherheit, 2001).

26. BHStA, LfN 45: Anton Micheler, *Was geht am Lech vor?* May 2, 1950; Kraus, *Bis zum letzten Wildwasser,* 34–42; Bernd Schmidt, "Der Lechausbau als Eingriff in eine Naturlandschaft" (thesis, Augsburg University, 1980), 43–52.

27. BHStA, LfN 45: Frickhinger and Ruess to MInn, May 24, 1949; Otto Kraus, "Naturschutz—ein Mahnruf 1958" (1958), in *Zerstörung der Natur,* 15–26; BHStA, LfN 47: Micheler to LfN, June 20, 1961.

28. Bayerische Wasserkraftwerke AG, *50 Jahre BAWAG,* 7, 24–25, 58; Bayerische Wasserkraftwerke AG, "Nochmals: 'Füssen ist Seestadt geworden,'" *Bayerische Staatszeitung,* Oct. 23, 1954; BHStA, StK 13773: *Stellungnahme zu dem Einspruch des Bayerischen Ministerrates gegen den Bau der Lechstaustufe 6 bei Schongau durch die Bayerischen Wasserkraftwerke AG,* July 8, 1955.

29. Otto Kraus, "Der Lech, der Forggensee und die BAWAG," *Blätter für Naturschutz*

34.1–2 (1954): 39–50; BHStA, StK 13773: Bund Naturschutz in Bayern to Hoegner, May 25, 1955; Kraus, *Bis zum letzten Wildwasser*, 40.

30. For a detailed discussion of the tension between nature conservation and mass tourism, see John Urry, "The Tourist Gaze and the Environment," in *Consuming Places*, John Urry (London: Routledge, 1995), 173–92.

31. See, e.g., BHStA, LfN 44: Kraus to Höfler, June 21, 1956; LfN 62: Regierungsstelle für Naturschutz to LfN, Jan. 5, 1950; LfN 63: Kraus to MInn, Sept. 21, 1962; LfN 65: Schmid-Hayn to Süddeutsche Zeitung, July 21, 1968; "Naturschützer wollen Watzmann retten," *Süddeutsche Zeitung*, Mar. 20, 1968.

32. See, e.g., Leslie Stephen, *The Playground of Europe* (London: Longmans, Green, 1871); Hans Magnus Enzensberger, "Eine Theorie des Tourismus," in *Einzelheiten*, ed. Hans Magnus Enzensberger (Frankfurt: Suhrkamp, 1962), 147–68. See also Walter Kiefl, "'Ich danke dir, daß ich anders bin als die anderen': Über das Phärisäertum der Kritik am Massentourismus," *Universitas* 48.7 (1993): 655–72; Christoph Hennig, "Trottel unterwegs: Zur Geschichte des Anti-Tourismus," *Gruppendynamik* 27 (1996): 7–11; and David Crouch and Nina Lübbren, introduction to *Visual Culture and Tourism*, ed. David Crouch and Nina Lübbren (Oxford: Berg, 2003), 1–22, esp. 5.

33. Manfred Pils, *Berg frei: 100 Jahre Naturfreunde* (Wien: Verlag für Gesellschaftskritik, 1994); Rainer Amstädter, *Der Alpinismus: Kultur—Organisation—Politik* (Vienna: WUV-Universitätsverlag, 1996); Helmuth Zebhauser, "Zwischen Idylle und Tummelplatz: Eine Ideengeschichte im neuen Alpinen Museum in München auf der Praterinsel," *Berg '97. Alpenvereinsjahrbuch* 121 (1997): 170–200; Judith Baumgartner, "Licht, Luft, Sonne, Bergwelt, Wandern und Baden als Sehnsuchtsziele," in *Die Lebensreform: Entwürfe zur Neugestaltung von Leben und Kunst um 1900: Band I*, ed. Kai Buchholz et al. (Darmstadt: Verlag Häusser, 2001), 407–10; Dagmar Günther, *Wandern und Sozialismus: Zur Geschichte des Touristenvereins "Die Naturfreunde" im Kaiserreich und in der Weimarer Republik* (Hamburg: Kovac, 2003).

34. In Germany, works by Jost Krippendorf (*Die Landschaftsfresser: Tourismus und Erholungslandschaft—Verderben oder Segen?* [Bern and Stuttgart: Hallwag, 1975]) and by Robert Jungk ("Wieviel Touristen pro Hektar Strand?" *Geo* 10 [1980]: 154–56) were of particular importance. See Dieter Kramer, "Tourismus ist mehr als nur 'ökonomische Transaktion,'" in *Reisebriefe: Sanfter Tourismus—ein Schlagwort mehr?* ed. Gruppe Neues Reisen (Berlin: Gruppe Neues Reisen, 1988), 59–70.

35. Jörg Maier, ed., *Naturnaher Tourismus im Alpenraum: Möglichkeiten und Grenzen* (Bayreuth: Universität Bayreuth, Lehrstuhl für Wirtschaftsgeographie und Regionalplanung, 1986); Gruppe Neues Reisen, ed., *Reisebriefe: Sanfter Tourismus—ein Schlagwort mehr?* (Berlin: Gruppe Neues Reisen, 1988); Peter Haßlacher, ed., *Sanfter Tourismus: Theorie und Praxis: Markierungen für die weitere Diskussion* (Innsbruck: ÖAV, 1989); Jürgen Hasse and Frauke Schumacher, *Sanfter Tourismus: Über ein konstruktives Verhältnis von Tourismus, Freizeit und Umweltschutz: Ein konzeptioneller Rahmen für die Verwaltung* (Bunderhee: Verlag für Umweltforschung, 1990), 11–12; J. Büchenhauer, *Die Alpen: Gefährdeter Lebensraum im Gebirge* (Cologne: Aulis-Verlag, 1996); Torsten Kirstges and Michael Lück, eds., *Umweltverträglicher Tourismus: Fallstudien zur Entwicklung und Umsetzung Sanfter Tourismuskonzepte* (Messkirch: Gmeiner, 2001).

36. BHStA, LfN 62–65; Otto Kraus, "Probleme um Seilbahnen" (1955), in *Zerstörung der Natur*, 211–15.

37. "Today, nature is one of our most important cultural possessions, however contradictory this may sound. One day, when the forty-hour week is being introduced, urban dwellers will escape to nature on a much grander scale than is already the case" (Wolfgang Engelhardt, "Alarm bei den Naturfreunden," *Landsberger Nachrichten*, December 1, 1955). See also Hubert Weinzierl, *Die große Wende im Naturschutz* (Munich: BLV-Verlagsgesellschaft, 1970).

38. BHStA, LfN 37: Kraus to Löw, Jan. 13, 1950; LfN 46: Micheler to mayor of Burggen, Nov. 27, 1959; Kraus, *Bis zum letzten Wildwasser*, 21, 34.

39. "This last section of primeval landscape on the Lech has to be preserved for scientists, artists, native friends of nature, users of collapsible boats, and strangers enjoying and enriching themselves on magnificent scenes of nature" (BHStA, LfN 45: Anton Micheler, *Was geht am Lech vor?* May 2, 1950).

40. Otto Kraus, "Leidensweg eines berühmten Naturschutzgebietes: Die Pupplinger Au bei Wolfratshausen, Obb." (1960), in *Zerstörung der Natur*, 194.

41. "This new jewel in the circle of Upper Bavarian natural beauties [i.e., the Forggensee] is developing into an outstanding center of attraction for tourism" (Haimerl, *Speicherkraftwerk Rosshaupten*, 16). See also "Streitobjekt Wendelsteinkraftwerk," *Süddeutsche Zeitung*, Apr. 1–2, 1950.

42. Haimerl, *Speicherkraftwerk Rosshaupten*, 16.

43. BHStA, LfN 46: Hilger memorandum, Dec. 1, 1962. See also "Wurde Landrat Dr. Hilger eine Million angeboten?" *Schongauer Nachrichten*, June 28, 1962, 5. This article quotes residents in the vicinity of a potential reservoir on the Lech as follows: "We profit little from a few nature lovers hiking in the Litzau with their backpacks. However, as soon as the major reservoir is finished, omnibuses are going to arrive here! Not to mention the benefits for the extension of our streets and roads."

44. Bayerisches Staatsministerium für Landesentwicklung und Umweltfragen, ed., *Forggensee: Erschliessung und Gewässergüte* (Munich: Bayerisches Staatsministerium für Landesentwicklung und Umweltfragen, 1971); BHStA, LfN 46: Kraus to MInn, Sept. 6, 1960; LfN 47: Lehrer- und Lehrerinnenverein München to Bayerischer Rundfunk, Nov. 7, 1962; Schmidt, "Der Lechausbau als Eingriff," 47.

45. According to John Urry, "to look individually or collectively upon aspects of landscape or townscape which are distinctive, which signify an experience which contrasts with everyday experience" is central to tourist consumption ("The Consumption of Tourism," in *Consuming Places*, Urry, 129–40, esp. 132; see also John Urry, *The Tourist Gaze: Leisure and Travel in Contemporary Societies* [London: Sage, 1990]).

46. Urry, "The Consumption of Tourism," 137–38.

47. Erich Steingräber, *Zweitausend Jahre europäische Landschaftsmalerei* (Munich: Hirmer Verlag, 1985), 326. See also Norbert Schneider, *Geschichte der Landschaftsmalerei: Vom Spätmittelalter bis zur Romantik* (Darmstadt: Primus Verlag, 1999), 7–13; Klaus Weschenfelder and Urs Roeber, eds., *Wasser, Wolken, Licht und Steine: Die Entdeckung der Landschaft in der europäischen Malerei um 1800* (Koblenz: Edition Braus, 2002), 83.

48. The sublime was well established as an aesthetic category by the third century

AD. In his treatise *A Philosophical Enquiry into the Origin of Our Ideas of the Sublime and Beautiful* (1757), Edmund Burke defined as possible sources of the sublime anything capable of evoking awe, pain, or horror—emotions that, in the absence of actual peril, are transformed into an experience of the sublime. On the sublime in Romantic painting, see Weschenfelder and Roeber, eds., *Wasser, Wolken, Licht und Steine*, 83.

49. Weschenfelder and Roeber, eds., *Wasser, Wolken, Licht und Steine*, 107.

50. Horst Ludwig, *Münchner Malerei im 19. Jahrhundert* (Munich: Hirmer Verlag, 1978); Steingräber, *Zweitausend Jahre*; Rolf Toman, ed., *Klassizismus und Romantik: Architektur, Skulptur, Malerei, Zeichnung 1750–1848* (Cologne: Könemann, 2000).

51. Crouch and Lübbren, introduction to *Visual Culture and Tourism*, 4–5. See also Antonia Dinnebier, *Die Innenwelt der Aussenwelt: Die schöne "Landschaft" als gesellschaftstheoretisches Problem* (Berlin: Technische Universität, Universitäts-Bibliothek, Abteilung Publikationen, 1996).

52. Reinhard Falter, "Die Gemeinden an der oberen Isar und ihr Verhältnis zum Fluss," in *Wahrnehmung, Bewusstsein, Identifikation: Umweltprobleme und Umweltschutz als Triebfedern regionaler Entwicklung*, ed. Kerstin Kretschmer and Norman Fuchsloch (Freiberg: Technische Universität Bergakademie, 2003), 131–54.

53. Painters who realistically depicted pre-Alpine rivers include Carl Rottmann, Johann Georg von Dillis, Joseph Wenglein, Otto Strützel, Wilhelm Scheuchzel, and Max Joseph Wagenbauer. For more on these figures, see Bernhard Setzwein, *An den Ufern der Isar: Ein bayerischer Fluss und seine Geschichte* (Munich and Berlin: Koehler und Amelang, 1993); Ludwig, *Münchner Malerei*; Horst Ludwig, *Von Adam bis Zügel: Bilder einer süddeutschen Privatsammlung* (Munich: Hirmer, 2001).

54. Schmidt, "Der Lechausbau als Eingriff," 14–15, 33–34. For photographs depicting the Lech as it was before the construction of hydroelectric power plants, see Heinz Fischer, "Der alte Lech," in *Der Lech: Wandel einer Wildflusslandschaft*, ed. Norbert Müller and Kurt R. Schmidt (Augsburg: Stadt Augsburg, Referat Umwelt und Kommunales, Amt für Grünordnung und Naturschutz, 1991), 37–58.

55. A good example of a river generally regarded as Romantic in appearance would be the middle section of the Rhine, with its medieval ruins and rich panoramas. For discussion of the Rhine's Romantic character, see Klaus Honnef, Klaus Weschenfelder, and Irene Haberland, eds., *Vom Zauber des Rheins ergriffen . . . : Zur Entdeckung der Rheinlandschaft* (Munich: Klinkhardt und Biermann, 1992). For depictions of other major German rivers, see Bärbel Hedinger, *Die Elbe malerisch gesehen* (Hamburg: Christians, 1992); Herwig Würtz, Ludwig Neunlinger, and Friedrich Schembor, *An der schönen blauen Donau: Donaupanoramen und -führer im Laufe der Zeit* (Vienna: Wien Kultur, 1995); Weschenfelder and Roeber, eds., *Wasser, Wolken, Licht und Sterne*.

56. Bayerische Wasserkraftwerke AG, "Nochmals: 'Füssen ist Seestadt geworden,'" *Bayerische Staatszeitung*, Oct. 23, 1954; Haimerl, *Speicherkraftwerk Rosshaupten*, 16; Bayerische Wasserkraftwerke AG, ed., *Der Lech und der Lechausbau* (Munich: Bayerische Wasserkraftwerke AG, 1974), 7.

57. Urry, "The Consumption of Tourism," 140; A. Bernatz, *Stellungnahme zur Broschüre "Die Partnachklamm ist in Gefahr": Eine sachliche und zeitgemäße Beurteilung des Projektes* KRAFTWERK WERDENFELS, 1949; StAM, Landratsamt Landsberg 193681: Bay-

erische Wasserkraftwerke AG, ed., *Der Speicher Rosshaupten und seine Bedeutung für den Ausbau des bayerischen Lech: Überreicht zur Tagfahrt des Wirtschaftsausschusses des Bayerischen Landtages zum Lech am 4. Mai 1950* (Munich: Bayerische Wasserkraftwerke AG, 1950), 3; Oscar Wolfbauer, "Projekt Kochelsee: Badeparadies für 30000," *Münchner Merkur,* May 27–28, 1970, 6.

58. Haimerl, *Speicherkraftwerk Rosshaupten,* 16.

59. Ibid. See also "Lechausbau der Bayerische Wasserkraftwerke AG bringt Strom und Arbeit," *Süddeutsche Zeitung* (special issue), Feb. 1951; Bayerische Wasserkraftwerke AG, *Der Lech und der Lechausbau,* 42–43.

60. In his article "Der Bach in der Landschaft" (1957), for example, Kraus simply argues that a valley through which an intact river flows is generally regarded as beautiful (*Zerstörung der Natur,* 136–44).

61. Wilhelm Jakob, "Der Forggensee bei Füssen," *Gemeinschaft und Politik* 3 (1955): 3–16, esp. 4. One of his contemporaries observed, "The times are long gone since one would have sounded the alarm hysterically because of such a construction and its impact on the scenery. We know that these wounds scar over, scar over beautifully like the work of a skilled surgeon" (Josef Martin Bauer, "Inn-Werk Neuötting wächst aus dem Boden," *Münchner Merkur,* Dec. 14, 1949, 5). See also Korbinian Lechner, "Sylvenstein und Roßhaupten," *Münchner Merkur,* May 27–29, 1950, 10; "Deutenhausen umspült schon der neue See," *Münchner Merkur,* July 31, 1953; "Das Allgäu erhielt einen neuen See," *Süddeutsche Zeitung,* May 22–23, 1954; A. Karbe, "Die Verzögerung des Energieausbaus kostet Millionen," *Bayerische Staatszeitung* 11 (1957); and Werner Glogauer, *Ein Alpenfluss—Der Lech* (Seebruck: Heering-Verlag, 1965), 21–22.

62. Alpine rivers today rank among the Central European habitats most substantially altered by humans over the past hundred years. For more on this topic, see Norbert Müller, "Veränderungen alpiner Wildflusslandschaft in Mitteleuropa unter dem Einfluss des Menschen," in *Der Lech: Wandel einer Wildflusslandschaft,* ed. Norbert Müller and Kurt R. Schmidt (Augsburg: Stadt Augsburg, Referat Umwelt und Kommunales, Amt für Grünordnung und Naturschutz, 1991), 10–29.

63. Kraus, "Leidensweg eines berühmten Naturschutzgebietes," 180–94. Among those areas that were protected from development were the Ammer, Partnach, Breitach, Weissbach, Wutach, and Weltenburg Danube Gorges; the waterfalls at Hölltobel and Tatzelwurm; the Soiern and Waging Lakes; and sections of the Lech, Iller, Tiroler Ache, Saalach, Wertach, and especially the Isar Rivers (Kraus, *Bis zum letzten Wildwasser,* 36).

64. Bergmeier, *Umweltgeschichte der Boomjahre,* 159–60, 166–67.

65. The term *major* is used for facilities of more than regional importance. Minor hydroelectric power plants mainly supply energy for the producer itself, while medium-sized stations fall between these two categories.

66. Examples of major hydroelectric power projects realized in the 1940s and 1950s in Bavaria are the translocation of the Rißbach River into the Walchensee in 1947–1949, the construction of the Sylvenstein reservoir on the Isar in 1954–1959, and the building of the Jochenstein power plant on the Danube in 1952–1955 (BHStA, LfN 37–40).

67. BHStA, LfN 38; StK 14652.

68. The Breitach Gorge near Oberstdorf, for example, which was also discussed as a potential location for a hydroelectric power plant, attracted 260,000 visitors in 1957 alone (Kraus, *Bis zum letzten Wildwasser*, 21).

69. BHStA, LfN 45: Huber to Oberste Baubehörde, Feb. 23, 1949; Town Council Füssen to Rural District Office Füssen, Feb. 23, 1949; Joachim Murat, "Es geht um die Partnachklamm," *Süddeutsche Sonntagspost* 46 (1949): 5–7; Hans Ehrhardt, "Der Füssener Lech: Ökologie und Naturschutz," in *Lebensraum Füssener Lech—Eine kleine Heimatkunde*, ed. Peter Nasemann (Füssen: Deutscher Alpenverein, Sektion Füssen, 1994), 43–60, esp. 58–59.

70. An example would be the case of the Forggensee (or Forggen Reservoir) on the Lech. See "Dörfer sollen überflutet werden," *Münchner Merkur*, Nov. 22, 1949; Hans Wüst, "'Waffenstillstand' am Lech," *Süddeutsche Zeitung*, June 19, 1951; and Theodor Strobl and Andreas Rimböck, "Der Forggensee am Lech—vom Energiespeicher zum Erlebnispark?" *Wasserwirtschaft* 90.9 (2000): 444–47.

71. The events at Waginger Lake in Upper Bavaria can be regarded as exemplary in this respect (BHStA, StK 17020, LfN 37, 38).

72. Examples would be the shifting of the waters of the Rissbach into the Walchensee, the destruction of the Illasberg Gorge at the River Lech, or the construction of the Forggen Reservoir near Füssen. See BHStA, StK 13775: Gröbner to Bayerischer Landtag, Feb. 24, 1947; Lettner petition, Mar. 12, 1947; BHStA, LfN 45: Kraus to Hofmann, July 16, 1950; Hofmann to Kraus, Sept. 11, 1950; BHStA, StK 13773: Seifert to Feneberg, Sept. 5, 1955; "Dörfer sollen überflutet werden," *Münchner Merkur*, Nov. 22, 1949; Will Berthold, "Still ruht der See—die Bauern streiten," *Münchner Illustrierte*, July 10, 1954; Falter, "Die Gemeinden an der oberen Isar," 131–54.

Chapter 10: Viewing the Gilded Age River

1. Peter Bacon Hales, "American Views and the Romance of Modernization," in *Photography in Nineteenth-Century America*, ed. Martha A. Sandweiss (Fort Worth: Amon Carter Museum, 1991), 205–57, esp. 206. See also Miles Orvell, "Viewing the Landscape," in *American Photography*, by Miles Orvell (Oxford: Oxford University Press, 2003), 39–60.

2. John F. Sears, *Sacred Places: American Tourist Attractions in the Nineteenth Century* (New York: Oxford University Press, 1989).

3. Alan Trachtenberg, *The Incorporation of America: Culture and Society in the Gilded Age* (New York: Hill and Wang, 1982); Robert Wiebe, *The Search for Order, 1877–1920* (New York: Hill and Wang, 1967).

4. Joan M. Schwartz, "*The Geography Lesson*: Photographs and the Construction of Imaginative Geographies," *Journal of Historical Geography* 22.1 (1996): 16–45; Susan Sontag, *On Photography* (New York: Anchor Books, 1977), 9–10; John Urry, *The Tourist Gaze: Leisure and Travel in Contemporary Societies* (London: Sage, 1990), 138–40; Dean MacCannell, *The Tourist: A New Theory of the Leisure Class*, 2nd ed. (New York: Schocken, 1989), 45. In describing the construction of an "imaginative geography," I am drawing on Edward Said's influential formulation in *Orientalism* (first published in 1978), in which

he shows how representations of other places and landscapes reflect the desires, fantasies, and preoccupations of their authors, as well as the fields of power between them and their subjects; see Edward S. Said, *Orientalism* (New York: Vintage Books, 1994). See also Derek Gregory, "Imaginative Geographies," *Progress in Human Geography* 19 (1995): 447–85.

5. Martha A. Sandweiss, *Print the Legend: Photography and the American West* (New Haven: Yale University Press, 2002); Rebecca Solnit, *River of Shadows: Eadweard Muybridge and the Technological Wild West* (New York: Viking, 2003); Weston Naef and James N. Wood, *Era of Exploration: The Rise of Landscape Photography in the American West, 1860–1885* (Boston: New York Graphic Society, 1975); Richard N. Masteller, "Western Views in Eastern Parlors: The Contribution of the Stereograph Photographer to the Conquest of the West," *Prospects* 6 (1981): 55–71; Peter Bacon Hales, *William Henry Jackson and the Transformation of the American Landscape* (Philadelphia: Temple University Press, 1988); Peter Palmquist, *Carleton E. Watkins, Photographer of the American West* (Albuquerque: University of New Mexico Press, 1983).

6. John Szarkowski, *The Photographer and the American Landscape* (New York: Museum of Modern Art, 1963), 3; Szarkowski, *Looking at Photographs: 100 Pictures from the Collection of the Museum of Modern Art* (New York: Museum of Modern Art, 1973), 44–45; Szarkowski, *American Landscapes: Photographs from the Collection of the Museum of Modern Art* (New York: Museum of Modern Art, 1981), 10; Martha Sandweiss, ed., *Photography in Nineteenth-Century America* (Fort Worth: Amon Carter Museum, 1991), 319; Tom Bamberger, "A Sense of Place," in *H. H. Bennett: A Sense of Place,* ed. Tom Bamberger and Terrance Marvel (Milwaukee: Milwaukee Art Museum, 1992), 11.

7. Joel Snyder, "Territorial Photography," in *Landscape and Power,* ed. W. J. T. Mitchell (Chicago: University of Chicago Press, 1994), 175–202.

8. Wiebe, *Search for Order,* 11–43; Hales, "American Views," 206.

9. Hales, "American Views," 206.

10. John Jakle, *The Tourist: Travel in Twentieth-Century North America* (Lincoln: University of Nebraska Press, 1985); Sears, *Sacred Places;* John Taylor, "The Alphabetic Universe: Photography and the Picturesque Landscape," in *Reading Landscape: Country-City-Capital,* ed. Simon Pugh (Manchester: Manchester University Press, 1990), 177–96; Patrick McGreevy, *Imagining Niagara: The Meaning and Making of Niagara Falls* (Amherst: University of Massachusetts Press, 1994); Dona Brown, *Inventing New England: Regional Tourism in the Nineteenth Century* (Washington, DC: Smithsonian Institution Press, 1995); Patricia Jason, *Wild Things: Nature, Culture, and Tourism in Ontario, 1790–1914* (Toronto: University of Toronto Press, 1995); John Towner, *An Historical Geography of Recreation and Tourism in the Western World, 1540–1940* (New York: John Wiley, 1996).

11. The problem of art and its relation to mechanical reproduction receives its classic treatment in Walter Benjamin, "The Work of Art in the Age of Mechanical Reproduction" (1936), in *Illuminations,* ed. Hannah Arendt (New York: Schocken, 1969), 217–52.

12. Sara Rath, *Pioneer Photographer: Wisconsin's H. H. Bennett* (Madison: Tamarack Press, 1979).

13. H. H. Bennett, diary entry, Feb. 25, 1866, Folder 3, Box 5; Diary/Receipts/Expenses,

1867, Folder 4, Box 5, both in H. H. Bennett Papers, Wisconsin Historical Society, Madison (hereafter HHBP), emphasis in original.

14. H. H. Bennett, letter to father, Feb. 5, 1867, Folder 9; Bennett, letter to father and mother, Oct. 20, 1867, Folder 10, both in Box 1, HHBP.

15. For a discussion of how Bennett's career path paralleled that of other landscape photographers during this period, see Frank Henry Goodyear, "Constructing a National Landscape: Photography and Tourism in Nineteenth-Century America" (Ph.D. diss., University of Texas at Austin, 1998).

16. P. Donan, *The Tourists' Wonderland: Containing a Brief Description of the Chicago, Milwaukee, and St. Paul Railway Together with Interesting General Descriptive Matter Pertaining to the Country Traversed by This Line and Its Connections* (Chicago: R. R. Donnelley and Sons, 1884), 23; Frank H. Taylor, *Through to St. Paul and Minneapolis in 1881: Random Notes from the Diary of a Man in Search of the West* (Chicago: Chicago, Milwaukee, and St. Paul Railway, 1881), 17.

17. *Wisconsin Mirror,* Apr. 7, 1872; Wisconsin Central Railroad, *Pen and Camera* (Milwaukee: Louis Eckstein, 1890). Such a connection between landscape photographers and railroad-driven resort development is well documented for the Far West. Carleton Watkins, Alfred A. Hart, Charles Roscoe Savage, F. Jay Haynes, and Alfred J. Russell are only the best known among the western landscape artists whose work helped promote railroad expansion. See Anne Farrar Hyde, *An American Vision: Far Western Landscape and National Culture, 1820–1920* (New York: New York University Press, 1990).

18. Bennett's letters indicate that he answered thousands upon thousands of queries regarding hotels and boat tours. On occasion and for special visitors, he would guide parties along the river.

19. *Wisconsin Mirror,* Apr. 7, 1872; P. Donan, *The Dells of the Wisconsin, Fully Illustrated* (Chicago: Rollins, 1879), 7.

20. H. H. Bennett, letter to J. E. Porter, Jan. 1, 1887, Folders 1–5, Box 4, HHBP.

21. H. H. Bennett Financial Ledgers, 1870–1889, Box 30, HHBP.

22. Hales, *William Henry Jackson.*

23. William H. Metcalf, "A Plea for the Stereoscope," *Photographic Times and American Photographer* 8 (Jan. 1883): 12–13.

24. Thomas Southall, "White Mountain Stereographs and the Development of a Collective Vision," in *Points of View: The Stereograph in America—A Cultural History,* ed. Edward W. Earle (Rochester, NY: Visual Studies Workshop, 1979), 97–108.

25. A well-known nickname for the Wisconsin, "America's hardest-working river," appears on state historical markers and in popular books such as Nels Akerlund's *Our Wisconsin River: Border to Border* (Rockford, IL: Pamacheyon Publishing, 1997).

26. Szarkowski, *The Photographer and the American Landscape,* 4.

27. Szarkowski, *American Landscapes,* 10. During an early excursion on the river, Bennett exhibited a sublime sensibility that failed to find its way into his later photographs: "Went to the Dells. Water the highest it's been for a great many years. It's terrible, awful, sublime, majestic and grand" (H. H. Bennett, diary entry for Apr. 23, 1866, Folder 3, Box 5, HHBP). This response was in many respects typical for visual artists of his day. On the sublime as an aesthetic category, see Marjorie Hope Nicolson, *Mountain Gloom*

and Mountain Glory: The Development of the Aesthetics of the Infinite (1959; Seattle: University of Washington Press, 1997).

28. Malcolm Andrews, *The Search for the Picturesque: Landscape Aesthetics and Tourism in Britain, 1760–1800* (Stanford: Stanford University Press, 1989); Barbara Novak, *Nature and Culture: American Landscape and Painting, 1825–1875* (New York: Oxford University Press, 1995); and Hyde, *An American Vision*.

29. W. W. Howe, "American Victorianism as a Culture," *American Quarterly* 27.4 (1975): 507–32; Helen Lefkowitz Horowitz, *Culture and the City: Cultural Philanthropy in Chicago from the 1880s to 1917* (Chicago: University of Chicago Press, 1976); Stow Persons, *The Decline of American Gentility* (New York: Columbia University Press, 1973); John Kasson, *Amusing the Million: Coney Island at the Turn of the Century* (New York: Hill and Wang, 1978).

30. James Maitland, *The Golden Northwest* (Chicago: Rollins, 1879), 37. More generally, see Sears, *Sacred Places*.

31. James Gilbert, *Perfect Cities: Chicago's Utopias of 1893* (Chicago: University of Chicago Press, 1991), 1–22.

32. A. C. Bennett, "Wisconsin Pioneer in Photography," *Wisconsin Magazine of History* 22.3 (1939): 268–79, also available at http://www.wisconsinhistory.org/wmh/articles/bennett.asp

33. Alan Trachtenberg, "Naming the View," in *Reading American Photographs: Images as History, Mathew Brady to Walker Evans* (New York: Hill and Wang, 1989), 119–63.

34. It is interesting to note that at the time that Bennett was photographing the Wisconsin Dells, just twenty miles downriver in Portage, Wisconsin, Frederick Jackson Turner was making his own set of landscape studies that would eventually form the basis for his famous frontier thesis, laid out in chapter 1 of *The Frontier in American History* (New York: Henry Holt, 1921).

35. Steven Hoelscher, "A Pretty Strange Place: Nineteenth-Century Scenic Tourism in the Dells," in *Wisconsin Land and Life*, ed. Robert C. Ostergren and Thomas R. Vale (Madison: University of Wisconsin Press, 1997), 424–49.

36. E. L. Wilson, "Our Picture," *Philadelphia Photographer* (May 1875): 137. See also Barbara Novak, "Landscape Permuted: From Painting to Photography," in *Photography in Print: Writings from 1816 to the Present*, ed. Vicki Goldberg (1981; Albuquerque: University of New Mexico Press, 1988), 171–79.

37. William Cronon, "Telling Tales on Canvas: Landscapes of Frontier Change," in *Discovered Lands, Invented Pasts: Transforming Visions of the American West*, ed. J. D. Prown (New Haven: Yale University Press, 1992), 37–88, esp. 81. For a similar argument, see Angela Miller, *The Empire of the Eye: Landscape Representation and American Cultural Politics, 1825–1875* (Ithaca, NY: Cornell University Press, 1993).

38. Hales, "American Views," 232. Bennett himself made frequent reference to "scenery" in his voluminous correspondence as well as in his most important view catalogue, "Wanderings among the Wonders and Beauties of Western Scenery," 1883, Folder 18, Box 25, HHBP. For a more general discussion of the landscape concept, see Denis Cosgrove, *Social Formation and Symbolic Landscape* (1984; Madison: University of Wisconsin Press, 1998).

39. Beginning with Robert F. Berkhofer Jr., scholars have long pointed out that once the threat of Native American resistance was eliminated, images of native peoples were routinely appropriated by their conquerors for a wide variety of political, economic, and cultural ends. See Berkhofer's pioneering study, *The White Man's Indian: Images of the American Indian from Columbus to the Present* (New York: Vintage Books, 1978). For a more recent discussion, see Philip J. Deloria, *Playing Indian* (New Haven: Yale University Press, 1998).

40. P. Donan, for one, marvels at the speed with which "fashionable tourism" displaced the "savage world": "Gay yachting and rowing parties now skim the mirror-like smoothness of lakes . . . which not many moons ago were only stirred by the prow of the Sioux or Winnebago birch-bark canoe" (Donan, *The Tourists' Wonderland*, 7). For a detailed analysis of Bennett's Ho-Chunk photographs, see Steven Hoelscher, "Viewing Indians: Native Encounters with Power, Tourism, and the Camera," *American Indian Culture and Research Journal* 27.4 (2003): 1–51.

41. Bennett arranged with the John Arpin Lumber Company of Grand Rapids for himself and his son Ashley to accompany the logging crew on one of their trips downriver. The two joined the fleet in Kilbourn City and traveled for eight days and more than a hundred miles before returning home. Although Bennett was prepared to pay for his and his son's passage, Bennett's financial records indicate that Arpin received one set of the series instead of a cash payment for the trip; see Bennett to D. J. Arpin, Oct. 18, 1886, Folder 1, Box 4; also Folder 6, Box 19, both in HHBP.

42. Although the anarchy associated with early lumber rafting has not been fully explored, it was widely covered in contemporary news reports. For just a few of the many articles on this topic, see *Wood County Reporter,* Apr. 16, 1859; *Madison Weekly Argus and Democrat,* Apr. 26, 1859; and *Milwaukee Daily Sentinel,* Apr. 14, 1859. For more on what labor historians call the Great Upheaval of 1886, see David Montgomery, *Workers' Control in America: Studies in the History of Work, Technology, and Labor Struggles* (Cambridge: Cambridge University Press, 1979). Gail Bederman's *Manliness and Civilization: A Cultural History of Gender and Race in the United States, 1880–1917* (Chicago: University of Chicago Press, 1996) also discusses changing work, racial, and gender relations in the period.

43. Rosalind Krauss, "Photography's Discursive Spaces," in *The Originality of the Avant-Garde and Other Modernist Myths,* by Rosalind Krauss (Cambridge, MA: MIT Press, 1985), 131–50. This is a central point in Sandweiss, *Print the Legend.*

44. Mike Crang, "Picturing Practices: Research through the Tourist Gaze," *Progress in Human Geography* 21.3 (1997): 359–73.

45. H. H. Bennett to George C. Lusseden, Jan. 3, 1899, reel 2, HHBP.

46. A. C. Bennett, "A Wisconsin Pioneer in Photography," *Wisconsin Magazine of History* 22.3 (1938–39): 268–79; S. J. Squire, "Accounting for Cultural Meanings: The Interface between Geography and Tourism Studies," *Progress in Human Geography* 18.1 (1994): 1–16.

47. H. H. Bennett to John Bennett, July 28, 1902, reel 3, HHBP. See also Hoelscher, "A Pretty Strange Place." As Peter Bacon Hales demonstrates in "American Views," the same

factors that diminished Bennett's profits and ultimately his professional status also affected view photographers more generally.

48. Cashbooks, 1870–89, 1890–1910, Folders 5–9, Box 35, and Folders 1–4, Box 36, HHBP.

49. H. Woody, "International Postcards: Their History, Production, and Distribution, circa 1895–1915," in *Delivering Views: Distant Cultures in Early Postcards*, ed. C. M. Geary and V. L. Webb (Washington, DC: Smithsonian Institution Press, 1998), 13–46.

50. J. Ruby, "Images of Rural America: View Photographs and Picture Postcards," *History of Photography* 12.4 (1988): 327–42; Hales, "American Views," 238.

51. This evolution is detailed in Hoelscher, "Viewing Indians." Later in the twentieth century, view photography and tourism merged in the powerful work of Ansel Adams. In his dramatic images of nature as a pre-human, Edenic wilderness, Adams's photographs depart significantly from Bennett's Victorian vision of nature as a genteel playground.

52. The full impact of the Wisconsin Dells dam project has not yet been analyzed, but landscape architect John Nolen was one of the first to see its hazards. In his report on the future state park system of Wisconsin, Nolen emphasized that the Dells are "Wisconsin's most characteristic and precious possession in the form of natural scenery. They are unique. For picturesqueness, romantic scenery, for alternative suggestions of mystery and majesty, the features of interest are numerous and varied." But Nolen was quick to caution that "this is what the Dells are today. What they will be when the dam now under construction is completed, when the water is raised permanently eighteen to twenty feet above the present, is not easy to say" (John Nolen, *State Parks for Wisconsin*, Report of the State Park Board of Wisconsin [1909], 27–29). The Dells, of course, were not made a state park. For his part, H. H. Bennett understood perfectly the effect that the impending dam construction would have on the river he knew so well. As one of the only people in Kilbourn City to challenge the dam—in a manner that anticipated John Muir's outcry, seven years later, against the damming of Yosemite's Hetch Hetchy Valley—Bennett wondered, "Why should such beauty spots [as the Dells] be destroyed or even injured as has been by building dams and cutting trees here for the profit of perhaps a few men when the good God made it that all people for all times might enjoy the beauties?" (H. H. Bennett, letter to Julia Lapham, Jan. 1, 1906, reel 5, HHBP). Bennett died a little more than a year after the Southern Wisconsin Power Company broke ground for the dam. For a triumphalist account of the power industry, see Forrest McDonald, *Let There Be Light: The Electric Utility Industry in Wisconsin, 1881–1955* (Madison: American History Research Center, 1957).

53. Yi-Fu Tuan, "Language and the Making of Place: A Narrative-Descriptive Approach," *Annals of the Association of American Geographers* 81 (1991): 684–96.

Contributors

Isabelle Backouche is an associate professor of history at the École des Hautes Études en Sciences Sociales (EHESS, Paris) and at the École Polytechnique (Palaiseau). She is also on the editorial board of *Genèses: Sciences sociales et histoire*. She is the author of *La trace du fleuve: La Seine et Paris (1750–1850)* (2000) and, more recently, "Entrer dans Paris par voie d'eau: Usages et urbanisation du bassin de La Villette au XIXe siècle" in *Entrer en ville*, edited by Françoise Michaud-Fréjaville et al. (2006).

David Blackbourn, Coolidge Professor of History and director of the Center for European Studies at Harvard University, is the author of *Class, Religion and Local Politics in Wilhelmine Germany: The Centre Party in Württemberg before 1914* (1980); *The Peculiarities of German History: Bourgeois Society and Politics in Nineteenth-Century Germany* (1984), with Geoff Eley; *Populists and Patricians: Essays in Modern German History* (1987); *Marpingen* (1993); *History of Germany 1780–1918: The Long Nineteenth Century* (1997, 2003); and *The Conquest of Nature: Water, Landscape, and the Making of Modern Germany* (2006).

Charles E. Closmann, assistant professor of history at the University of North Florida, holds a doctorate from the University of Houston and is a former Research Fellow of the German Historical Institute in Washington, DC. His essays have appeared in the *Journal of Urban History*, *Hamburger Wirtschaftschronik*, and the *Bulletin of the German Historical Institute*; he is also the editor of *War and the Environment: Contexts and Consequences of Military Destruction in the Modern Age*, forthcoming in 2008.

Timothy M. Collins, associate dean for research at the University of Wolverhampton's School of Art and Design and director of the Black Country Centre for Art and Design Research and Experimentation in the UK, is also a distinguished research fellow of the STUDIO for Creative Inquiry at Carnegie Mellon University. Among his recent publications are "Art Nature and Aesthetics in the Post-Industrial Public Realm" in *Healing Natures,*

Repairing Relationships: New Perspectives on Restoring Ecological Spaces and Consciousness, edited by Robert L. France (2007), and "An Ecological Context" in *New Practices/New Pedagogies: A Reader*, edited by Malcolm Miles (2005).

Jacky Girel is a research engineer in the laboratory for alpine ecology of the Centre National de la Recherche Scientifique, at Joseph Fourier University in Grenoble, France. His current research focuses on how historical changes in land use have affected plant biodiversity. He is the author and co-author of scientific studies on numerous topics, such as environmental management and the ecological integrity of alpine floodplains since the development of civil-engineering works.

Ute Hasenöhrl is a doctoral candidate at the Free University in Berlin, Germany, and has also worked as a scientific assistant at the Social Science Research Centre in Berlin and as a subject specialist at the Hessian State Library in Wiesbaden. She is the author of *Zivilgesellschaft, Gemeinwohl und Kollektivgüter* (2005) and the editor of a collection titled *Historismus und Moderne: Blick zurück nach vorn?* (2007). Her dissertation focuses on the nature conservation movement in Bavaria from 1945 to 1980.

Steven Hoelscher is associate professor of American Studies and Geography at the University of Texas at Austin. He is the author of *Heritage on Stage* (1998), of articles published in *American Quarterly*, the *Annals of the Association of American Geographers*, the *Geographical Review*, and the *American Indian Culture and Research Journal*. His book, *Picturing Indians: Photographic Encounters and Tourist Fantasies in H. H. Bennett's Wisconsin Dells*, is forthcoming from the University of Wisconsin Press.

Thomas Lekan is associate professor of history and faculty associate in the School of the Environment at the University of South Carolina–Columbia. He is the author of *Imagining the Nation in Nature: Landscape Preservation and German Identity, 1885–1945* (2004). With Thomas Zeller, he co-edited *Germany's Nature: Cultural Landscapes and Environmental History* (2005). He has received fellowships from both the German Historical Institute of Washington, DC, and the American Council of Learned Societies for his current project, *Sublime Consumption: Nature Tourism and the Making of the German Leisure Class, 1880–1980*.

Christof Mauch holds the chair in American history and transatlantic relations at the University of Munich. From 1999 to 2007 he was the director of the German Historical Institute in Washington, DC. He has authored, edited, and co-edited many books in the fields of German history and literature, U.S. history, and environmental history. Among his most recent publications are *Nature in German History* (2004), *Berlin–Washington, 1800–2000: Capital Cities, Cultural Representation, and National Identities* (2005), *Shades of Green: Environmental Activism Around the Globe* (2006), and *The World Beyond the Windshield: Roads and Landscapes in the United States and Europe* (2008).

Edward K. Muller is professor of history and director of the urban-studies program at the University of Pittsburgh. He is co-author, with John F. Bauman, of *Before Renaissance: Planning in Pittsburgh, 1889–1943* (2006); editor of *DeVoto's West: History, Conservation, and the Public Good* (2005); and co-editor, with Thomas F. McIlwraith, of the second edition of *North America: The Historical Geography of a Changing Continent* (2001).

Joel A. Tarr is the Richard S. Caliguiri University Professor of History and Policy at Carnegie Mellon University. He is the author, with Clay McShane, of *The Horse in the City: Living Machines in the Nineteenth Century* (2007) and *The Search for the Ultimate Sink: Urban Pollution in Historical Perspective* (1996). He is also editor of *Devastation and Renewal: An Environmental History of Pittsburgh and Its Region* (2004).

Dorothy Zeisler-Vralsted, professor of government and vice president for student affairs at Eastern Washington University, is also the author of several articles on water development and has contributed essays on the same topic to other edited collections. At present, she is researching a comparative study of the historical development of the Volga and Mississippi rivers from the nineteenth century to the present.

Thomas Zeller, associate professor of history at the University of Maryland, College Park, is also a former research fellow and visiting research fellow at the German Historical Institute of Washington, DC. He is the author of *Driving Germany: The Landscape of the German Autobahn, 1930–1970* (2007) and the coeditor of three recent essay collections: *How Green Were the Nazis? Nature, Environment, and Nation in the Third Reich* (2005), *Germany's Nature: Cultural Landscapes and Environmental History* (2005), and *The World Beyond the Windshield: Roads and Landscapes in the United States and Europe* (2007).

Index

Adams, Ansel, 152
Adenauer, Konrad, 128
Afghanistan, 77
Africa, 4
agriculture, 14–15, 17, 19, 23, 72, 78, 83, 86–88; and pollution, 101–2, 104, 110; and river transportation, 63, 68–69, 75, 76
Ahr River, 123
Aire River, 90–92
Albertville, 87
Allegheny County Sanitary Authority (ALCOSAN), 53
Allegheny River, 8, 41–62
Alliance for the Protection of German Waters (*Vereinigung deutscher Gewässerschutz, VDG*), 105, 121,123, 129, 131
Aluminum Company of America (ALCOA), 57
American Rivers (conservation organization), 75
Anthony's (journal), 161
Arc River, 83
Army Corps of Engineers (United States), 3, 46, 52, 59, 63, 66, 69–71
Arndt, Ernst Moritz, 125, 130
Arno, 1, 11
artwork. *See* rivers: in popular culture
Aswan High Dam, 12, 21
Atlantic Ocean, 33
Austria, 139
Auxerre (France), 28
Azov Sea, 68

Backouche, Isabelle, 60
Baden, 3, 16
Balke, Siegfried, 121
Baltic Sea, 12, 68
Baltimore, 57
Banvard, John, 65–66
Basel, 17, 110–11
Bassin de La Villette, 30
Bastille, 36–37

Bavaria, 22, 138–48
Bayerische Wasserkraftwerke AG (BAWAG), 139–42, 145–46
Beaver River, 43
Beck, M.B., 98
Becker, August, 18
Bennett, Henry Hamilton: and The Bennett Studio, 9, 151–71
Bergmeier, Monika, 111
Berlin, 13, 14, 15, 20, 100
Bernoulli, Daniel: and Bernoulli's Principle, 14
Biedermeier art, 144–45
Big Structures, Large Processes, Huge Comparisons (Tilly), 12
Bildzeitung, 122
Bingen, 18
Bingham, George Caleb, 68
Birmann, Peter, 16
Black Forest, 117
Black Sea, 68
Bleiloch dam, 21
Bonn, 123
Bordeaux, 32
Boston, 49, 52
Bradford, 91, 93, 95
Brazil, 4
Breisach, 111, 117, 119
Breitkreutz, Ernst, 15
bridges, 33, 38, 44, 180n2
Bridges over the Rhine (Noth), 130
Brüggemeier, Franz-Josef, 89, 104
Brullée, Jean-Pierre, 37
Bundesrat, 122
Bundestag, 106
Bund Naturschutz (conservation organization), 138

Calder River, 90–93
Camera's Story of Raftsman's Life on the Wisconsin, The (Bennett), 163–66
Canada, 39–40

223

Canal du Midi, 3
canals, 2–3, 5, 20, 40, 43, 62, 104, 117; France, 30, 33, 35, 37; Rhine, 104, 117; United States, 43. *See also specific canals*
Carson, Rachel, 125
Caspian Sea, 65, 68
Catherine II of Russia, 74
Central Commission for Rhine Shipping, 111
Chaney, Sandra, 111, 124
Charles Emmanuel III, King of Sardinia, 82
Château Trompette, 32
chemical plants. *See* pollution
Chicago, 9, 149, 160, 163, 164
China, 1, 2, 5, 77
Choron, Louis, 85
Cioc, Mark, 6–7, 100
Civil War (United States), 3, 43, 68, 150
coal industry: in England, 90–92; in Germany, 20, 99–102, 104, 106, 110, 115–16, 130, 139; in the United States, 43, 44, 50, 53, 55, 59
Cold War, 90
Coleridge, Samuel, 115
Cologne, 115
Colorado River, 6, 149, 151
Columbia River, 6, 25, 151
Columbia Valley Authority, 71, 76
Congress (United States), 3, 46, 52
conservation, 22–23, 53, 55, 70, 111–14, 117, 124–29, 137, 139–48
Corbet, Charles-Louis, 36
Cronon, William, 62, 161

dams, 4–5, 7, 12, 18–24, 77, 140–42, 149; locks and dams systems, 3, 43, 46–47, 62, 63, 66–67, 69–72, 75
Dana, Charles A., 66
Dante Alighieri, 1
Danube, 12
deforestation, 13, 20, 81, 129
de Gaulle, Charles, 130
Demoll, Reinhard, 117, 118, 121, 124, 127–28
de Saussuer, Horace Bénédict, 83
Deutsche Alpenverein, 138, 142
Deutsche Zeitung, 121
De Wailly, Charles, 35–36
dikes, 11, 13, 14, 15, 16, 39, 79, 82–88
Dimitrov, 74
disease. *See* public health and disease
displacement of citizens, 4, 10, 21, 30, 74
Dixon, Samuel G., 49–50
Dmitriev, Ivan, 67
Dnieper River dam, 72

Dominick, Raymond, 111, 118, 124
Dominik, Hans, 19
Dortmund, 100, 102
Douglas, Mary, 113, 129
drainage systems, 79, 86–87, 95
draught, 7, 52, 95, 120; and water rationing, 122
drinking water. *See* water supply
Duisburg, 100, 105, 115
Dunbar, William P., 92
Düsseldorf, 99, 106, 122
Düsseldorfer Nachrichten, 122

East India Company, 33
École Nationale des Pontes et ChausÊees, 34, 38, 81, 85, 86
Ecological Master Plan for the Rhine (Salmon 2000), 132
ecosystems. *See* wildlife habitats
Egypt, 1
Eliot, T.S., 11
Ely, Christopher, 67
Emerson, Ralph Waldo, 155
employment: and hydraulic projects, 70, 73, 83
Emscher Cooperative, 102–5, 107, 116
Emscher River, 99, 101–5, 107, 116
End of Nature, The (McKibben), 6
End of Summer on the Volga (Savrasov), 67
England, 3, 60, 89–98, 107–9
Entenfang, 17
environmental activists, 12, 90, 98, 111–12, 117–19, 121–25
environmental history, 5–10, 12, 15, 23–25, 62, 111–13
Erzgebirge, 19
Eschbach Dam, 19
Essen, 100, 101, 102, 104
Euler, Leonhard, 14
Euphrates, 1, 11, 25
Europe, 5, 7
European Economic Community (EEC), 107, 109
European Union (EU), 98, 107, 109
European Water Protection Federation, 131
Evening on the Volga (Levitan), 67

Farnam, Henry, 66
Federal Water Law (WHG) (Germany), 106
Fighting League (activist group), 124
Fillmore, Millard, 66
Finch, John, 97
Fink, Mike, 66

Firin, Syemyen, 73
fish, 13, 23, 65, 75, 120, 121; in England, 92, 95, 96; fishing industry, 65, 104, 123, 137; fish kills, 25, 53, 101, 123, 132; migration, 17, 18, 22; return to formerly polluted rivers, 58; in the Rhine, 18, 24, 114, 117, 132; salmon, 24, 111, 123; in the United States, 43, 50
Fitz-James, Jacques, 80
floods, 4, 11, 13, 15, 41–42, 44, 51, 85, 117; control and prevention, 3, 21, 30, 39, 51–55, 81–82, 149. *See also* dams
Florence, 1
Forggensee reservoir, 144, 145–46
Four Quartets (Eliot), 11
Fradkin, Philip L., 6
France, 3, 17, 19, 26–40, 79–82, 85–87, 114, 130–32; French Revolution 17, 27, 36, 79, 82, 87. *See also specific cities or geographic locale*
François, Antoine, 82
Frankfurt, 121
Frankland, Edwin, 94
Frederick the Great, 13–16
Friends of the Riverfront, 58
Füssen, 146

Ganges, 11
Garella, Francesco-Luigi, 80
Garella, Napoleon, 82–83
Garner, J.H., 96
Garonne River, 32
Geertz, Clifford, 12
Gelsenkirchen, 100
Genscher, Hans Dietrich, 106
German Association of Gas and Water Experts, 119
German Council for Agriculture, 102
German Society for Public Health, 100
Germany, 12; Federal Republic, 105–9, 110–36, 139–44; German Democratic Republic, 15, 123, 139; Nazi period, 23, 112–14, 126–29, 132; Weimar Republic, 112. *See also* Prussia; *specific cities and geographic locale*
Giraud, Pierre, 35–36
Goethe, Johann Wolfgang von, 18
Goldgrund, 17
Gorky, Maxim, 64
Grand Canal (China), 2
Great Depression (United States), 70, 76
Green Party, 111, 135–36
Grenoble, 87
groundwater, 47, 115, 117, 140, 146

Grzimek, Bernhard, 136
Gulf of Mexico, 65

Haerlem, Simon Leonhard, 14
Hales, Peter Bacon, 150
Halifax, England, 93
Hall, Charles, 70
Hamlin, Christopher, 94
Hardin, Blaine, 6
Haussmann, Baron, 37–38
Haynes, F. Jay, 151
Hazen, Allen, 49–50
Heimat, 112–14, 126–36
Heimat: Bearbeitung und Gestaltung (Demoll), 127
Heimatschutz, 126, 132, 136
Heinz, H.J., 51
Hennig, Richard, 19
Heraclitus, 11
Heuer, Carl, 14
Honnef, 115
Hoover Dam, 4, 21, 71
Hoover, Herbert, 70
Hornsmann, Erich, 121
Hôtel-Dieu, 35–36
Hudson River, 149
Hudson River School (artists), 160
Humber (river and tidal estuary), 90, 98
Hurricane Katrina, 4
Hurricane Rita, 4
hydroelectricity, 4–5, 18, 20, 71–72, 74, 76–77, 124, 137–44, 147–48, 149, 170

If We Did Not Have Water (Hornsmann), 121
Île de la Cité, 28
Île des Cygnes, 28, 33, 35
Île Feydeau, 32
Île Louviers, 35, 37
Île Saint-Louis, 35
Illinois River, 75
Imhoff, Karl, 116; "Imhoff tanks", 103, 116
Intze, Otto, 19, 20, 21
India, 5, 77
industrial pollution. *See* pollution
industrial use of water. *See* water supply
industrialization, 44–47, 50–52, 56, 58, 61, 64, 72, 74, 121, 137–39, 143, 149; the Industrial Revolution, 90, 100, 144
International Commission for the Protection of the Rhine against Pollution (Rhine Commission), 9, 115, 131–35
International Working Group of the Waterworks of the Rhine Basin (IAWR), 131

irrigation, 5, 19, 23, 71
Isar River, 145, 147
Isère River, 79–88
Itaipú Dam, 4
Itasca, Lake, 65
Izaak Walton League, 70

Jaag, Otto, 131
Jackson, William Henry, 151, 155
Jakob, Wilhelm, 146
Jakobsson, Eva 6
Jolly Flatboatmen (Bingham), 68

Kalkar, 119
Kama River, 65, 74
Karamzin, Nikolai, 67
Karlsruhe, 16, 17
Käsebier, Gertrude, 170
Kiskiminetas River, 43
Klosterkemper, Horst, 129
Koblenz, 18
Kraus, Otto, 124, 127
Krupp, 100
Kuibyshev Reservoir, 65; Kuibyshev dam, 74
Kun, Gleb, 73, 74

Lachine Canal, Montréal, 40
La Dhuis River, 39
La Vanne River, 39
Lawrence, David, 54
Lech River, 9, 138–42, 145–46
Leeds, 91, 95
legislation and regulations: in England, 90, 92–98; in France, 26; in Germany, 90, 102, 106–9, 111, 122, 126, 132; in the United States, 42, 50, 53, 58, 70
Leibl School, 145
Lenin, Vladimir, 76
Letschin, 16
levees, 4, 40
Levitan, Isaak, 67
Lewis, Henry, 66
Lier, Adolf, 145
Life on the Mississippi (Twain), 66
Lilienthal, David, 64, 71
Lippe River, 99
locks and dams. *See* dams: locks and dams systems; Mississippi River: locks and dams
Loire River, 32
London, 33, 39; fog disaster of 1952, 113
Longfellow, Henry Wadsworth, 155
Lorenz, Konrad, 136

lumber industry, 2, 9, 69, 163–66, 171
Luxembourg, 115, 132
Lyon, 40

Machiavelli, Niccolò, 11
Mainz, 18
"Man and Nature" (VDG article), 129
Man as Member and Shaper of Nature (Thienemann), 120
Mangin, Charles, 36
Mannheim, 115–16
Manstein, Bodo, 118, 121
Marckolsheim, 119
Marcus Aurelius, 11
McKibben, Bill, 6
Mead, Elwood, 71
Mecklenburg, 19
media: coverage of water issues, 54, 111, 117, 122, 124, 146
Mellon, Richard King, 54
Melun (France), 28
Merced River, 151
Michalsky, Werner, 15
Middle Rhine Gorge, 126–27
Minnesota, 64–65; Minneapolis, 65, 66, 69
Mississippi River, 3, 8, 16, 40, 42, 64–67, 163; locks and dams, 63, 67, 68–72, 75–77
Missouri River, 65
Möbius, Karl, 120
Monde, Le, 110
Monongahela River, 8, 41–62
Montereau, 28
Montréal, 40
Morgenstern, Christian, 145
Moscow, 63, 68
Moscow-Volga Canal, 63–65, 67–68, 72–77
Mulheim, 101
Munich, 145, 147

Nantes, 32
Native Americans, 43, 150, 155, 163, 170
NATO, 130
Nature in Danger (film), 127
navigation. *See* rivers: and transportation
Nekrasov, Nikolai, 67, 68
Netherlands, 2, 40, 114, 115, 131, 132
Newcastle-on-Tyne, 60
New Deal, 64, 70–72, 76–77
New Orleans, 3
New York City, 49
New York Daily Times, 66
New York Tribune, 66
Niagara Falls, 150

INDEX

Nidd River, 90
Nile River, 1, 11, 12
North Rhine-Westphalia, 105, 115–16, 127, 129, 141
North Sea, 12
Noth, Ernst, 130
nuclear power plants, 124; protests against, 111, 119, 124; and thermal pollution, 110, 118–19
nuclear weapons, 118
Nye, David, 19

Oder River, 12–16, 23, 24; Oder Marshes, 13–16
Ohio River, 8, 41–62, 65
Oka River, 65
Old Man River (journal), 70
Olmsted, Frederick Law, Jr., 52, 55
Organic Machine, The (White), 6
Ortega y Gasset, José, 128
O'Sullivan, Timothy, 151
Ottawa River, 39–40
Our Achievements (journal), 64
Ouse River, 90

Palloy, Pierre-François, 36–37
Panama Canal, 70
Panorama of the Mississippi (Banvard), 65
Paraguay, 4
Paris, 7, 26–40, 60–62
Parliament (English), 92–93, 95–98, 109
Parliament (French), 33–34
Partnach Gorge, 147
Pascal, Blaise, 23
Passemant, Claude Siméon, 33
Pennsylvania Environmental Council, 58
Pennsylvania Main Line, 43
Pennsylvania Railroad, 44
Périer, J.C., 36
Perronet, Jean-Rodolphe, 34
pesticides, 110–11, 123, 125
Peter the Great, Tsar, 68
Pettenkofer, Max von, 117
Philadelphia Photographer, 155, 161
Phönix, 100
Photographic Times, 155
photography: of riverscapes, 150–71
Pinchot, Gifford, 52
Pisa, 11
Pittsburgh, 8, 41–62
Place Royale (Paris), 36
pollution, 28, 39, 42, 47, 70, 110–35; chemical plants, 110, 115–16, 122, 131–32; industrial waste, 50, 55, 58, 75, 90–97, 101–9, 116, 124, 149, 175; mining, 50, 53, 55, 101, 104, 115, 116, 131; thermal pollution, 18, 110, 118–19, 131. *See also* coal industry; legislation and regulations; public health and disease; sewage
Pont au Change, 30
Pont Neuf, 28
Pont Notre-Dame, 27, 30
Porter, Eliot, 152
Portland (Oregon), 57
Powell, John Wesley, 5
Poyet, Bernard, 35
"Preservation of Federal Waterways, The" (Klosterkemper), 129
Prince, The (Machiavelli), 11
Principles of Rational Agriculture (Thaer), 15
Prussia, 8, 13–16, 89–90, 100–105, 116
public health and disease, 31, 32, 34, 39, 42, 51, 53, 55, 89, 98, 100, 108; cholera, 39, 115; malaria, 15, 83–87; typhoid, 39, 48–49, 101
Purity and Danger (Douglas), 113

Quai de Brancas, 32
Québec City, 40

Raftsmen Playing Cards (Bingham), 68
railroads, 39, 44, 47, 66, 68–69, 154, 163
Ranke, Leopold von, 11
Ranstad, 40
Ratzel, Friedrich, 120
recreation. *See* rivers: recreation and tourism
reforestation, 20, 51–52, 59, 85
Régemorte, Louis, 34
Reich Nature Protection Law (1935), 126–27
Reichstag, 102, 116
Remscheid, 19
Repin, Ilya Efimovich, 67, 68
Report on the Lands of the Arid Regions of the United States (Powell), 5
reservoirs, 4, 7, 20, 21–22, 103–4, 139, 141, 143, 145–46, 148
Rheingold (Wagner), 17
Rheinische Stahlwerke, 100
Rhenish Society for Landscape Protection and Monument Conservation, 136
Rhine Commission. *See* International Commission for the Protection of the Rhine against Pollution
Rhine River, 1, 6, 12, 23–25, 66, 99; and hydraulic engineering, 16–18; pollution of, 3, 103, 104, 110–36

Rhine Action Plan (1987), 110, 132
Rhine Valley Action Committee, 119
Rhône River, 1, 40
Rideau Canal, 40
Riehl, Wilhelm Heinrich, 125
River and Harbor Act (US, 1930), 70
River Don, 95
River Lost, A (Harden), 6
rivers: and boating, 40, 54, 57; and ecotourism, 142–43; in popular culture, 65–68, 73, 74, 144–45, 150–71; recreation and tourism, 9, 35, 39, 44, 53, 61, 62, 66, 137–38, 141–51, 154–55, 158, 163, 171; and riverfront development, 40, 42, 52, 57; and transportation, 2–3, 17–18, 23, 60, 64, 65, 67–77; and transportation in France, 28, 30–40; and transportation in the United States, 41, 43–47, 52, 68–72, 149, 163
Rivers of Empire (Worster), 5
Rock Island Railroad, 66
Rome, 1
Roosevelt, Franklin Delano, 71, 76; and the New Deal, 64
Roosevelt, Theodore, 52, 69
Rouen, 28, 33
Ruhr Dam Association /Cooperative for a Clean Ruhr, 103–4, 106, 107–8, 109
Ruhr region, 8, 20, 89–90, 99–109, 115–16
Ruhr River, 19, 20, 99, 101, 120
Ruhr Valley Reservoirs Association, 20
Rybinsk Dam, 65, 74

Salmon and Trout Magazine, 95
Salpêtrière Hospital (Paris), 33, 35, 37
Salton Sea, 22
Saône River, 40
Samara River, 65
Sandoz chemical factory fire (1986), 111, 132, 136
sanitation. *See* public health and disease
Sauerland, 141
Savoy, 79–88
Savrasov, Aleksei, 67
Scandanavia, 12
Schleich, Eduard, 145
Schwab, Günther, 118, 124
Schwartz, Joan, 151
Scott, James C., 15
Seeger, Hendrik, 106
Seeing Like a State (Scott), 15
Seine, 7, 26–40, 60–62
Sens, 28
sewage: dispersion into rivers, 44, 53, 55, 89, 92–95, 101, 115, 129; sewage systems, 39, 48–50, 100; sewage treatment, 49, 54, 95–98, 114, 103–9, 115–16, 127. *See also* pollution; legislation and regulations
Sheail, John, 89, 93, 96, 108
Sheffield, 91, 95
Sheffield, Joseph, 66
Sheksna River, 65
Short, Alfred, 95
Sieg River, 111
Silent Spring (Carson), 125
Silesia, 20
Smith, Paul, 95
South America, 4
Soviet Union, 64, 71, 72–77, 118
Spiegel, Der, 122
Sprague, N.S., 47
St. Louis, 40, 65
Stalin, Joseph, 68; and the Five-Year Plan, 64, 72–74
steamboats, 66
steel industry, 58, 99–100, 106
Stieglitz, Alfred, 170
Strasbourg, 17
Sumarokov, Aleksandr, 67
Switzerland, 16, 87–88, 114, 132, 168
Syndicat de l'Isère et de l'Arc, 85
Szarkowski, John, 157

Tate Gallery, 39
Tennessee Valley Authority (TVA), 64, 71, 76–77
Thaer, Daniel Albrecht, 15
Thames, 1, 33, 39
Thames Conservancy, 97
Thatcher, Margaret, 98, 108
Thienemann, August, 22, 101, 120
Thoreau, Henry David, 66
Thorsheim, Peter, 113
Three Gorges Dam, 5
3 Rivers 2nd Nature (environmental organization), 58
Three Rivers Stadium, 55
Thyssen, August, 101
Tiber, 1
Tigris, 1
Tilly, Charles, 12
Toulouse, 32
tourism. *See* rivers: recreation and tourism
Touristenverein Die Naturfreunde (conservation organization), 142
Trachtenberg, Alan, 160
transportation. *See* rivers: and transporta-

INDEX 229

tion; rivers: and transportation in France; rivers: and transportation in the United States
Trent River, 90
Trinius, August, 23
Trotsky, Leon, 71
Tulla, Johann Gottfried, 3, 16–18, 19, 23
Twain, Mark, 66
Two Dimitrovs (Kun), 74
Tygart River, 42

Uekötter, Frank, 141
Umwelt, Umweltschutz, 111, 120–21, 125
United States, 7, 8, 64, 118; environmental movement, 125; Bureau of Reclamation, 4, 71; Fish and Wildlife Service, 59; hydraulic engineering, 2–5, 22 , 71; legislation related to water quality, 42, 52, 53, 58, 70,106; Public Works Administration, 70, 71. *See also* Army Corps of Engineers; *specific cities and geographic locale*
Upper Mississippi River Improvement Association, 69
urbanization, 115, 121, 143, 149
Urry, John, 144

Valday Hills, 65
Vetluga River, 65
Victor Amadeus III, King of Sardinia, 82
Victor Emmanuel, King of Italy, 82–83
Vistula River, 12, 23
Volga Boatmen, The (Repin), 67
Volga River, 8, 63–65, 67–68, 72–77
Volta River, dam, 21
Voyages dans les Alpes (de Saussure), 83

Waginger Lake, 147
Wagner, Richard, 17
Wakefield, 93–94
Wakefield Express, 94
Walchensee, 22, 147n66, 148n72
Walters, R.C.S., 95
Warthe River, 15
Water Act of 1973 (England), 97
Water Management Law (Germany, 1957), 132

water shortages, 94, 97, 101, 103, 108;
Wassernot (water emergency), 111, 120–24
water supply: and drinking water, 20, 23, 28, 35, 38–39, 41, 44, 47–49, 53, 63, 72, 75, 89, 92–94, 96–98, 105, 108–9, 110, 114, 115, 118, 122, 132; filtration/purification, 39, 49,114, 129; industrial and commercial use of, 23, 41, 44, 47, 72, 89, 92, 94, 97, 101, 103–6, 119. *See also* pollution
Wath-on-Deane, 94
Watkins, Carleton, 151
Welt Die, 122
Werdenfels region, 147
Western Pennsylvania Field Institute, 58
West Fork River, 42
Westphalia, 116
Wharfe River, 90–91
Whipple, George C., 49–50
White, Richard, 6, 25
White Sea Canal, 73
wildlife habitats, 25; degradation of, 4, 17, 42, 50, 62, 63, 75, 87, 118; reemergence of 58–59; wildlife refuges, 22, 70, 148. *See also* fish
Wilhelm II, Kaiser, 20
Willamette River, 151
Wisconsin, 69, 149–71
Wisconsin River and Dells, 9, 149–71
Wittfogel, Karl, 5
Wohl, Ellen E., 6
World Commission on Dams, 4
World League for the Defense of Life, 124
World War II, 8, 21, 54, 112–15, 135
Worms, 17, 115
Worster, Donald, 5, 23
Wupper, 19, 104
Wyhl power plant, 111, 119, 136

Yagoda, Genrikh, 73
Yangzi River, 1, 5
Yellowstone National Park, 150
Yezhov, Nikolai, 73
Yorkshire, England, 8, 89–98, 107–9
Youghiogheny River, 43
Yvette Aqueduct, 38